Robot Adventures in Python and C

Thomas Bräunl

Robot Adventures
in Python and C

 Springer

Thomas Bräunl
Engineering and Mathematical Sciences
The University of Western Australia
Perth, WA, Australia

ISBN 978-3-030-38899-7   ISBN 978-3-030-38897-3   (eBook)
https://doi.org/10.1007/978-3-030-38897-3

This Springer imprint is published by the registered company Springer Nature Switzerland AG
The registered company address is: Gewerbestrasse 11, 6330 Cham, Switzerland

PREFACE

Contrary to common belief, the three most important topics in robotics are *not* Mechanics, Electronics and Software. Instead, they are Software, Software and Software! Many years ago, I wrote the book *Embedded Robotics* that kept a balance between electronics and software, while also presenting some basic mechanics content. This new book, however, concentrates mainly on software for mobile robots. This is where the real challenge lies and where the real innovation is happening.

In this book we demonstrate how inexpensive mobile robots such as our *EyeBot* robots can be constructed by mounting a Raspberry Pi controller and camera onto a model car or some other simple mechanical drive system. And we introduce our *EyeSim* simulation system, which is freely available and can quite realistically simulate a variety of driving, swimming/diving and even walking robots. Our emphasis is on algorithm development, and we ensure that all software projects can run on the real robot hardware as well as on the simulation system. This means, we do not use any unrealistic simulation assumptions that would never work in the real world.

At The University of Western Australia, we found that students using EyeSim as a supplementary teaching tool in robotics greatly improved their learning rate and understanding of robotics concepts.

All software used in this book, including all example programs, can be downloaded from the links below. There are native applications for MacOS, Windows, Linux and Raspberry Pi.

EyeBot real robots: `http://robotics.ee.uwa.edu.au/eyebot/`
EyeSim simulation: `http://robotics.ee.uwa.edu.au/eyesim/`

In the following chapters, we will start with simple applications and move on to progressively more complex applications, from a small, simple driving robot to a full-size autonomous car.

Preface

This book contains source code for most of the problems presented. In order to keep some order, we use a color-coding scheme to distinguish:

- Python programs
- C/C++ programs
- SIM scripts
- Robot definition files
- Environment data files

Tasks and challenges at the end of each chapter will help to deepen the learned concepts and let readers use their creativity in writing robot programs.

I hope you will enjoy this book and have fun recreating and extending the applications presented – and then go on to create your own robotics world!

My special thanks go to the UWA students who implemented EyeSim and also wrote some of the example programs: Travis Povey, Joel Frewin, Michael Finn and Alexander Arnold. You have done a great job!

Thanks for proofreading of the manuscript go to Linda Barbour and the team at Springer-Verlag.

Perth, Australia, March 2020   Thomas Bräunl

CONTENTS

Contents

Contents

ROBOT HARDWARE

I n this book, we will talk about a number of fundamentally different mobile robots – from small basic driving robots, via autonomous submarines and legged robots, all the way to driverless cars. At the Robotics and Automation Lab at The University of Western Australia, we have developed the *EyeBot Family* (see Figure 1.1), a diverse group of mobile robots, including wheeled, tracked, legged, flying and underwater robots [Bräunl 2008][1]. Each robot has a camera as the main sensor and carries a touchscreen LCD as its user interface.

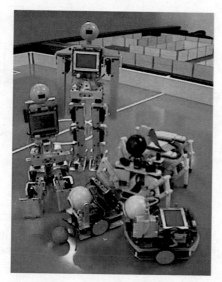

Figure 1.1: EyeBot mobile robot family

Robots are, in fact, closely linked to Embedded Systems. They comprise a (smaller or larger) on-board computer system, which is connected to actuators

---

[1]   T. Bräunl, *Embedded Robotics – Mobile Robot Design and Applications with Embedded Systems*, 3rd Ed., Springer-Verlag, Heidelberg, Berlin, 2008

© Springer Nature Switzerland AG 2020
T. Bräunl, *Robot Adventures in Python and C*, https://doi.org/10.1007/978-3-030-38897-3_1

and sensors. The computer system continuously reads sensor input to get information about the world surrounding it, and then reacts by sending commands to its actuators. Actuators are mostly electric motors, such as wheel motors or leg-servos, but could also be pneumatic or hydraulic actuators, solenoids, relays or solid state electronic switches.

1.1 Actuators

A driving robot's motion capability comes from a number of actuators – in most cases two electric motors. The simplest mechanical drive system one can imagine is called *differential drive*, which is a robot platform that has two separately controlled wheel motors attached to it (see Figure 1.2). If both motors are driven forward at equal speed, then the robot is driving forward in a straight line. Likewise, for driving backwards. If one motor is running faster than the other, e.g. the left motor is going faster than the right motor, then the robot will drive in a curve – in this case a right (or clockwise) curve. Finally, if one motor is going forward (e.g. the left) and the other motor backwards, then the robot is turning on the spot (here: clockwise).

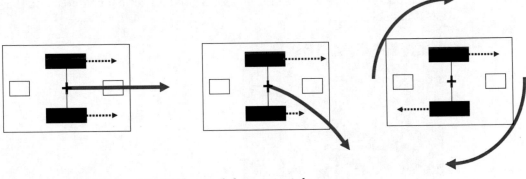

Figure 1.2: Differential drive principle

1.2 Sensors

The motion functionality is only one half of a mobile robot. The other half is sensing. Even for our simplest robots, we use three types of sensors. In order of complexity, these are shaft encoders, infrared distance sensors and digital cameras.

Shaft Encoder Shaft encoders are simple sensors that provide feedback to control the speed of a robot's motors (velocity control) as well as measure the distance the robot has travelled (position control). They count every small motor shaft rotation and translate it via the robot's kinematic formula into a change of its translational and rotational position (*pose*). Of course, this can only work if the robot is firmly on the ground and there is no wheel slip. But even then, there is always a small inaccuracy for each encoder "tick", so over a larger distance travelled, position and orientation derived from encoder values alone will become quite inaccurate and unreliable.

A slotted disk (Figure 1.3) is usually used that alternates in letting an infrared LED beam through and blocking it. While the wheel is rotating, this generates a rectangle signal (Figure 1.4).

Figure 1.3: Incremental encoder (left); motor with encapsulated gearbox and encoder (right)

A variation of this principle is to use a reflective disk of alternating white and black sectors (Siemens Star) in combination with an LED and a detector on the same side.

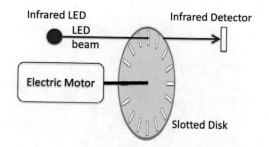

Figure 1.4: Encoder disk principle (left); encoder output signal for slow-fast-slow rotations (right)

Infrared PSD (position sensitive device) Infrared distance sensors are also known as Position Sensitive Devices (PSD). They emit a light beam invisible to the human eye and use the reflection from an object to calculate a distance value. Depending on where the reflected beam lands on a detector array, the closer or further away the object is (see Figure 1.5, right). Our EyeBot robots use at least three of these sensors, pointing to the front, left and right, to calculate the distances to walls or obstacles in these directions.

PSDs come in a variety of different shapes and forms (e.g. in Figure 1.5, left). They also have different interfaces, i.e. analog or digital output values.

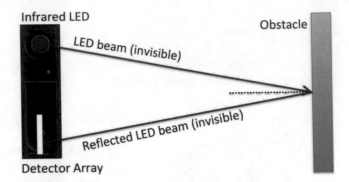

Figure 1.5: Sharp PSD sensor (left) and measurement principle (right)

Digital Camera A digital camera is a much more complex and powerful sensor than the ones mentioned before. It delivers millions of pixels per image frame, several times per second. At the very moderate VGA[2] resolution there are 640×480 pixels with 3 Bytes each at 25 Hertz (PAL[3]) or 30 Hertz (NTSC[4]), so over 23 MB/s for PAL and almost 28 MB/s for NTSC. Figure 1.6 shows our own *EyeCam* system (left) next to the Raspberry Pi camera module (right). The standard Raspberry Pi camera has a fixed non-changeable lens. If the field of view (or another camera parameter) does not suit your robot application, you can use a third-party camera with a replaceable board lens instead. A large variety of board lenses are available to suit most applications.

Figure 1.6: EyeCam M4 (left) and Raspberry Pi camera (right)

---

2   VGA: Virtual Graphics Adapter; an image resolution of 640×480 pixels, first introduced for the IBM PS/2 in 1987

3   PAL: Phase Alternating Line; the European analog TV standard with 625 lines at 25 frames (50 alternating half-frames) per second, matching a 50Hz mains power frequency

4   NTSC: National Television System Committee (also jokingly called *"never the same color"*); the North-American analog TV standard with 525 lines at 30 frames (60 alternating half-frames) per second, matching a 60Hz mains power frequency

As the sheer amount of data requires a powerful controller for processing, we usually work with a relatively low image resolution to maintain an overall processing speed of 10 fps (frames per second) or more. The images in Figure 1.7 have been taken at a resolution of 80×60 pixels – probably as low as one would like to go, but as you can see, nevertheless, with a rich amount of detail visible.

Figure 1.7: Sample images with a resolution of 80×60 pixels

1.3 User Interface

Although technically not really necessary, all of our robots carry a user interface in the form of a touchscreen display (see Figure 1.8). With this, the robot can display sensor and measurement results, and the user can enter commands of select parameters using soft buttons. This interface works with the physical LCD on the "back" of the real robot, in a laptop window via WiFi remote desktop from a PC or even in simulation.

 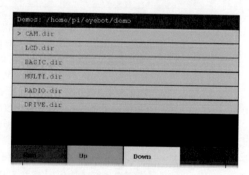

Figure 1.8: Real robot touchscreen (left) and remote desktop input (right)

1.4 Processor

Actuators and sensors have to be interfaced to an embedded controller. Our choice was a Raspberry Pi in combination with our own *EyeBot7* board for I/O and motor drivers (see Figure 1.9, right). It is based on an Atmel XMEGA A1U processor, similar to an Arduino, and links to the main controller via a USB line. The EyeBot7 robotics I/O-board provides a number of interfaces that the Raspberry Pi (in Figure 1.9, left) does not have:

- 4 H-bridge motor drivers with encoders
- 14 servo outputs
- 16 digital I/O lines
- 8 analog input lines

 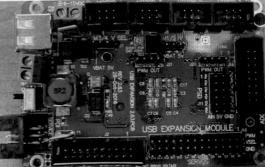

Figure 1.9: Raspberry Pi controller (left) and EyeBot M7 I/O-board (right)

1.5 Complete Robot

Putting it all together lets us build a complete robot. The differential drive system can be in the middle or in the back of the robot chassis with one or two passive caster wheels to prevent the robot falling over. We add PSD sensors facing front, left and right plus the camera looking to the front. The Raspberry Pi controller is sandwiched between the touchscreen on top and the EyeBot7 I/O-board below. The photos in Figure 1.10 show our compact *SoccerBot S4*, while Figure 1.11, left, shows the mechanically simpler but somewhat wider *EyeCart*.

The diagram in Figure 1.12 shows the principal hardware setup. Display, camera and high-level sensors (e.g. GPS, IMU, Lidar etc. with either USB or LAN connection) are directly linked to the Raspberry Pi controller. The EyeBot7 I/O-board builds the bridge to the drive motors, servos and low-level sensors. It communicates via USB with the Raspberry Pi.

An even simpler approach without a dedicated I/O-board is shown in Figure 1.13. When using a model car platform that has a built-in servo for steering

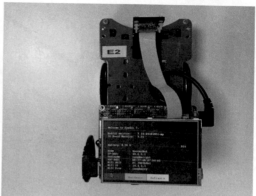

Figure 1.10: SoccerBot S4 robot

Figure 1.11: EyeCart robot (left with I/O controller) and simple robot chassis (right without I/O)

and a digital motor controller, then for driving we only need two PWM (pulse-width modulation) output lines, which are directly accessible on the Raspberry Pi controller. However, on the Raspberry Pi controller these are implemented in software, while the EyeBot7-I/O board has dedicated PWM hardware. Since we do not use a real-time operating system on the Raspberry Pi, there can be some noticeable jitter in the steering servo due to timing variations in the processing of different tasks on the controller. Although there will be similar variations for the drive motor control, these variations tend to matter less and are barely noticeable.

This approach can be used for building a cheap, minimal configuration driving platform, but of course it lacks the low-level sensors, especially the shaft encoders. It may not be suitable for precision driving, but the model car base does allow much higher driving speeds.

A similar approach can be used with a very basic robot chassis as shown in Figure 1.11, right. The link between the Raspberry Pi and the chassis with two

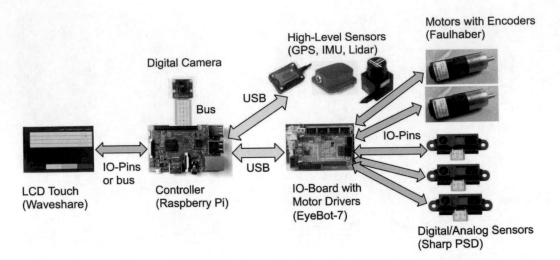

Figure 1.12: Robot system structure with I/O controller board

differential drive motors is a cheap motor driver board attached to the side of the Pi, while the camera is velcroed to the front. The whole robot is powered by a USB power bank. As before, this approach misses feedback from wheel encoders and distance data from infrared PSD sensors, which would require a low-level I/O board for interfacing.

For further details on drive mechanics, electronics hardware and system software see the EyeBot User Guide [Bräunl et al. 2018][5].

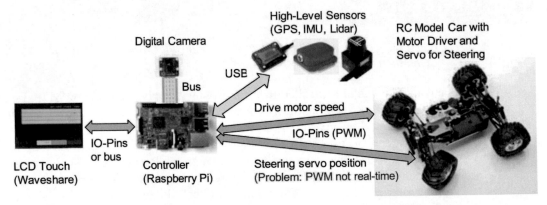

Figure 1.13: Robot system structure without I/O controller board

---

5   T. Bräunl, M. Pham, F. Hidalgo, R. Keat, H. Wahyu, *EyeBot 7 User Guide*, 2018, http://robotics.ee.uwa.edu.au/eyebot7/EyeBot7-UserGuide.pdf

1.6 Communication

Each robot is now an independent autonomous vehicle. However, it is highly desirable even for a single robot to have a wireless communication link to a laptop or desktop PC, in order to transfer programs to and data back from the robot. When using more than one robot, we would like to have a communication network that lets the robots communicate with each other.

Our network is based on the Raspberry Pi's built-in WiFi module. As a default, each robot has its own WiFi hotspot, so we can easily connect to it with a laptop, tablet or smartphone.

The default WiFi hotspot network name and password are
   PI_12345678 and *raspberry*

where the number is automatically derived from the Pi's MAC address, allowing the operation of several independent robots in the same room.

The default IP address for WiFi is easy to remember:
   10.1.1.1

The default username and password for our EyeBot-Raspian distribution are
   pi and *rasp*

When using a LAN cable to connect, it is the same username and password, with an equally simple default IP address:
   10.0.0.1

When using several robots as a group, their network setting can be changed to "slave", linking them to a common DHCP[6] WiFi router. That way, all robots can talk to each other directly, as well as to a non-robot base station, e.g. the operator's laptop computer.

1.7 Simulation

In the following chapters of this book we will frequently work with simulation. We should stress that we have created a very realistic approximation of real robot behavior. Robot programs can be directly transferred from the simulation to the real robot without having to change a single line of source code. There are no unrealistic assumptions, no virtual (wishful) sensors, no "cheats". Even the error settings are very realistic and can be adapted. There is no perfect world: a robot being told to drive 1.0m will always deviate a little bit (e.g. 0.99m or 1.01m) and sensor readings are not always 100% correct either – and our simulation environment reflects this.

We have re-implemented the EyeSim simulation system several times from scratch since the first version some 20 years ago. The latest version is called

---

[6]   A DHCP router (Dynamic Host Configuration Protocol) assigns each WiFi client a unique IP (Internet Protocol) address.

Figure 1.14: EyeSim-VR simulation of a robot scene

EyeSim-VR (Figure 1.14), which runs natively on MacOS, Windows and Linux, and even supports Virtual Reality (VR) devices from Oculus and HTC. EyeSim-VR was implemented by Travis Povey and Joel Frewin in 2018 and extended by Alexander Arnold and Michael Finn in 2019.

We decided to structure this book in a project-based fashion, where it does not really matter whether we work with real or simulated robots – algorithms and program code will be identical. We will start with simple robotics tasks and develop more complex applications step-by-step.

1.8 Tasks

- Configure your ideal mobile robot by selecting wheels, actuators (motors) and sensors, such as a camera, PSDs, etc.
- Make a spreadsheet showing part numbers, quantities, suppliers and cost.
- Make a CAD drawing of your robot design, considering dimensions of all components.
- Build it!

ROBOT SOFTWARE

2

E yeSim is a realistic mobile robot simulation system, developed by the Robotics & Automation Lab at The University of Western Australia, that is distributed free of charge. It supports a number of different robot types and sensors, and produces realistic, close-to-reality motion patterns. EyeSim robot simulation source code can be directly ported to the real physical EyeBot robots without changing a single line of code. Supported programming languages are Python, C and C++.

Included in EyeSim are various wheeled and tracked driving robots, omnidirectional robots with Mecanum wheels, various AUVs (autonomous underwater robots) and the legged robot *Starman*. More information is available in the EyeSim User Manual[1] and on the EyeSim web page (see Figure 2.1):

> http://robotics.ee.uwa.edu.au/eyesim/

 EyeSim VR - Unity Based EyeBot Simulator

* Now with submarines! *

Introduction

The EyeBot simulator with virtual reality *EyeSim VR* is a **multiple mobile robot simulator with VR functionality** based on game engine Unity 3D that allows experiments with the same unchanged EyeBot programs that run on the real robots. EyeSim VR is capable of simulating all major functionalities in RoBIOS-7, including:

- LCD Output/Key Input
- Camera output

Figure 2.1: EyeSim VR website

---

[1] EyeSim VR Team, *EyeSim User Manual*, 2018, robotics.ee.uwa.edu.au/eyesim/ftp/EyeSim-UserManual.pdf

Select the software package matching your operating system: MacOS, Windows or Linux. EyeSim has been implemented using the Unity 3D[2] games engine, which allows it to run natively on each of these platforms. Its built-in physics engine also makes all robot movements very realistic.

Additional system requirements are X11 libraries for MacOS and Windows, and our adapted version of *Cygwin* for Windows. Additional packages, such as OpenCV[3] for image processing, are optional.

2.1 Software Installation

Depending on your operating system, you need to install some auxiliary software as outlined in Figure 2.2. In addition, there are optional software packages available, such as OpenCV for high-end image processing.

Operating Systems	OS Version	Prerequisites
Windows	Windows 8.1, 10	Cygwin (EyeSim version), Xming
Mac OS	10.10.X	XQuartz
Linux	64 bit	–

Figure 2.2: EyeSim-VR system requirements

All additional software can be found at
    `http://robotics.ee.uwa.edu.au/eyesim/ftp/aux/`

The EyeSim packages are located at
    `http://robotics.ee.uwa.edu.au/eyesim/ftp/`

After installation, have a look at the EyeSim User Manual
    `http://robotics.ee.uwa.edu.au/eyesim/ftp/EyeSim-UserManual.pdf`

and also at the EyeBot User Guide (for the real physical robots)
    `http://robotics.ee.uwa.edu.au/eyebot7/EyeBot7-UserGuide.pdf` .

The key to robot control is the set of functions that form the *RoBIOS* API (Robot Basic IO System – Application Programmer Interface). This is included in the EyeBot User Guide (see Figure 2.3) or can be seen on the web at
    `http://robotics.ee.uwa.edu.au/eyebot7/Robios7.html` .
Figure 2.4 gives an overview of RoBIOS print functions; the complete list of RoBIOS functions is included in the Appendix of this book.

When you have EyeSim correctly installed, you should have a symbol on your task bar, such as the one on the left. Clicking on this will start the EyeSim simulator with a default environment in which a single robot has been placed (see Figure 2.5).

---

2   Unity 3D, https://unity3d.com
3   OpenCV is the Open Source Computer Vision Library, see OpenCV.org

**EyeSim VR
User's Manual**

EyeSim VR Team

November 3, 2017

[Revised November 13, 2018]

EyeBot 7 User Guide

Thomas Bräunl, Marcus Pham, Franco Hidalgo, Remi Keat, Hendra Wahyu
August 29, 2018

EyeBot 7 is the 2017 version of the EyeBot embedded controller for robotics applications. It is now based on a Raspberry Pi board with optional LCD display, linked via USB to the EyeBot7-IO board, which has hardware and software drivers for motors and digital or analog sensors. The board runs Raspian Linux with the RoBIOS user interface software on top and provides an extensive robotics library that allows the simple design of robot application programs in C using the RoBIOS API.

Link

http://robotics.ee.uwa.edu.au/eyebot7/

In the following, we will discuss each of these components separately.

Contents

1. EyeBot User Interface
2. RoBIOS Library
3. Hardware Description Table
4. EyeBot IO-Board
5. Building a Robot

Figure 2.3: EyeSim VR User's Manual and EyeBot 7 User Guide

RoBIOS-7 Library Functions

Version 7.1, Oct. 2018-- RoBIOS is the operating system for the EyeBot controller.
The following libraries are available for programming the EyeBot controller in C or C++.
Unless noted otherwise, return codes are 0 when successful and non-zero if an error has occurred.

In application source files include: #include "eyebot.h"
Compile application to include RoBIOS library: $gccarm myfile.c -o myfile.o

- LCD Output
- Key Input
- Camera
- Image Processing
- System Functions
- Timer
- USB/Serial
- Audio
- Distance Sensors
- Servos and Motors
- V-Omega Driving Interface
- Digital and Analog I/O
- IR Remote Control
- Radio Communication

- Multitasking
- Simulation

LCD Output

```
int LCDPrintf(const char *format, ...);        // Print string and arguments on LCD
int LCDSetPrintf(int row, int column, const char *format, ...); // Printf from given position
int LCDClear(void);                            // Clear the LCD display and display buffers
int LCDSetPos(int row, int column);            // Set cursor position in pixels for subsequent printf
int LCDGetPos(int *row, int *column);          // Read current cursor position
int LCDSetColor(COLOR fg, COLOR bg);           // Set color for subsequent printf
int LCDSetFont(int font, int variation);       // Set font for subsequent print operation
int LCDSetFontSize(int fontsize);              // Set font-size (7..18) for subsequent print operation
int LCDSetMode(int mode);                      // Set LCD Mode (0=default)

int LCDMenu(char *st1, char *st2, char *st3, char *st4); // Set menu entries for soft buttons
int LCDMenuI(int pos, char *string, COLOR fg, COLOR bg); // Set menu for i-th entry with color [1..4]
```

Figure 2.4: RoBIOS-7 API functions

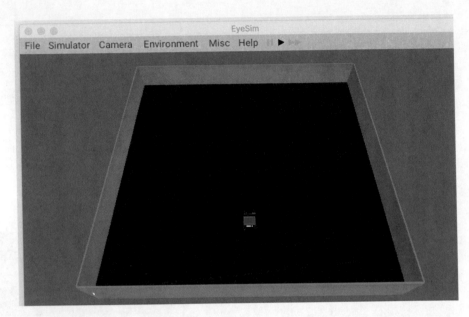

Figure 2.5: EyeSim default environment

This startup scenario can be changed via the File/Settings menu. You can have multiple robots, several objects, walls, markers, colors and textures, 3D terrain and even water. We will come back to this later.

2.2 First Steps in Python

For now, we will just try to get this robot moving. The relevant RoBIOS command for this is

```
int VWSetSpeed(int linSpeed, int angSpeed)
```

which we call the v-ω (v-omega) driving interface, as we can specify a velocity v and an angular velocity ω for the robot. If v is given a certain value and ω=0, the robot should be driving straight. With v=0 and ω set to a fixed value, the robot should be rotating in place, and if both v and ω get non-zero values, the robot will be driving a curve of some sort. Let's try this out!

The easiest start is using Python, but we will repeat this example in C as well. After starting EyeSim and having a robot ready as shown earlier in Figure 2.5, use a command window and type:

```
python3
```

or start a programming environment such as Thonny or PyCharm. After the Python command prompt, enter your robot program:

```
from eye import *
```

This will make all the RoBIOS API commands mentioned before available. Then type your first driving command, e.g.

```
VWSetSpeed(100,0)
```

which will set the robot on a straight path at 100mm/s with zero rotational speed. In the command window you will see the system dialog of Program 2.1.

Program 2.1: Programming robot from command line in Python

```
[tb-pro:~ tb$ python3
Python 3.6.5 (v3.6.5:f59c0932b4, Mar 28 2018, 05:52:31)
[GCC 4.2.1 Compatible Apple LLVM 6.0 (clang-600.0.57)] on darwin
Type "help", "copyright", "credits" or "license" for more information.
>>> from eye import *
>>> VWSetSpeed(100,0)
Connection established. Handshaking...
Handshake complete. Waiting for server ready...
Server ready. Beginning control.
0
>>>
```

At the same time in the EyeSim window you can see the robot driving forward. In fact, if we are not quick enough to stop it with

```
VWSetSpeed(0,0)
```

it will hit the back wall (Figure 2.6). No problem, it is only a simulated robot, so no damage has been done. You can click on it and move it back to the middle of the field. With the + and − keys you can also rotate the robot to any angle.

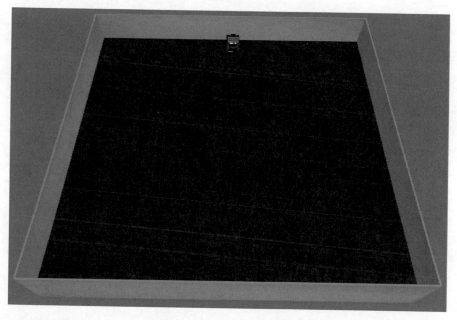

Figure 2.6: Robot controlled from Python commands

2.3 First Steps in C

Driving straight in C is done by the equivalent program in Program 2.2. But we then need to be quick to stop it if we want to avoid a wall collision.

Program 2.2: Robot controlled by program in C

```
1    #include "eyebot.h"
2    int main ()
3    { VWSetSpeed(100, 0);
4    }
```

The *include* statement makes the RoBIOS API available, and all C programs require a *main* function definition as the program start. The only statement in this program is the *VWSetSpeed* command. Note the semicolon after the statement and the curly brackets encapsulating the function definition. You will need a lot of these in C.

As C is compiled and not interpreted as with Python, we have to do the compilation step of our source program before we can run it. Although this may seem like unnecessary extra work, it is actually quite beneficial, as it checks the source code for errors and will report them – while Python will start executing a program and may then stop in the middle of the execution when it encounters an error.

Program 2.3: Compiling and executing a C program

```
●  ●  ●              📁 tmp — -bash — 52×26

tb-pro:tmp tb$ gccsim straight.c -o straight.x
tb-pro:tmp tb$ ./straight.x
Connection established. Handshaking...
Handshake complete. Waiting for server ready...
Server ready. Beginning control.
tb-pro:tmp tb$
```

Compilation in C is simple with the *gccsim* script we put together (see Program 2.3). The first parameter is the C source file and the "-o" option lets you specify the name of the binary output file:

```
gccsim straight.c -o straight.x
```

For all EyeSim example directories we provide so-called *Makefiles*, which greatly simplify compilation of C and C++ programs. With the correct *Makefile* in place, all you have to type for compilation is a single word:

```
make
```

Assuming we have started the EyeSim simulator already and have a robot waiting for commands, we can now run our program:

```
./straight.x
```

Linux always requires us to specify the directory when executing a command, so if the executable program *straight.x* is in our current directory, we have to prefix it with "./".

2.4 Driving a Square in Python

Let us take this up a notch and try drive the robot around a square. For this, we use two other API commands, which will drive straight or rotate in place but stop after the desired distance or angle, respectively. These commands are

```
int VWStraight(int dist, int lin_speed)
int VWTurn(int angle, int ang_speed)
```

Their parameters are in [mm] and [mm/s] for *VWStraight* and [degrees] and [degrees/s] for *VWTurn*. Program 2.4 is the complete Python program to drive one square.

Program 2.4: Driving a square in Python

```
1   from eye import *
2
3   for x in range (0,4):
4       VWStraight(300,500)
5       VWWait()
6       VWTurn(90, 100)
7       VWWait()
```

We repeat the sequence of driving straight four times then turning in place. The *for*-loop is doing this for us. Its parameters (0,4) indicate that the counter *x* will start at 0 and run while it is less than 4, so it will do the iterations 0, 1, 2, 3 – in total four runs, which is exactly what we are after.

The calls to *VWWait* after each driving command are necessary as the commands *VWStraight* and *VWTurn* will give control back to the program immediately and any subsequent driving command will override the previous one. So, if you forget to put *VWWait* into your program, the program will very quickly dash through all commands and only execute the very last *VWTurn* command. It will turn only 90° once and not move from its starting position.

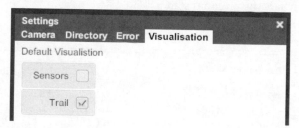

Figure 2.7: Visualization settings in EyeSim's menu

The simulator's menu command File / Settings / Visualization allows us to turn on visualization of the robot's path (Figure 2.7), which will make it easier for us to explain its movements. With this set, the square program will mark the robot's movements on the floor (see Figure 2.8).

Figure 2.8: Robot completing the square program

Finally, rather than typing the driving commands directly into the Python3 interpreter, we can put them into a file, e.g. *square.py*. We can then call the program with the command

```
python3 square.py
```

Or, to go one step further, we make the source file *square.py* executable (by changing its file permissions) and then add the name of the Python3 interpreter as the very first line:

```
#!/usr/bin/env python3
```

Now we can start the Python program even more easily, directly from the command line:

```
./square.py
```

2.5 Driving a Square in C or C++

Using C or C++ for the same driving task is not much more difficult, as shown in Program 2.5. The commands are identical, but C has its own syntax for including the EyeBot/RoBIOS library and needs a *main* function, which tells the system where the program starts.

Program 2.5: Square driving program in C

```
1    #include "eyebot.h"
2    int main()
3    { for (int i=0; i<4; i++)    // run 4 sides
4      { VWStraight(400, 300);    // drive straight 400mm
5        VWWait();                // wait until finished
6        VWTurn(90, 90);          // turn 90 degrees
7        VWWait();                // wait until finished
8      }
9    }
```

The *for*-loop is a bit wordier but does the same as in Python; it just runs the statement block between the inner curly brackets "{" and "}" four times. As before, *VWStraight* and *VWTurn* need to be followed by *VWWait* statements, to ensure that they are completely executed, before progressing to the next drive command.

2.6 SIM Scripts and Environment Files

We can now write, compile (C/C++ only) and execute a robot program. But when we are setting up more and more complex scenarios with structured driving environments, multiple robots and numerous objects in a scene, we could use some help so that we do not have to manually place all components again and again. This can be achieved by using a ".sim" script file like the one shown in Program 2.6.

Program 2.6: SIM script for a simple world with a single robot placement

```
1    # Default Environment
2    world rectangle.maz
3
4    # Robot placement
5    S4 1500 300 90   square.py
```

Besides the comments (starting with "#"), there are only two items in this *SIM* script: the *world* command selects a file that describes the driving environment, and the *S4* command places an S4-type robot into the environment at the specified (x,y)-coordinates (1500mm in x and 300mm in y) and rotation angle (90°). The executable file is our Python program *square.py* – but you could also replace this with *square.x* for a C/C++ binary file.

There are many more things that can be done with a *SIM* script, e.g. adding a line such as

```
settings VIS TRACE
```

that will automatically activate visualization of the robot's infrared distance sensors (*VIS*) and draw its driving path onto the floor (*TRACE*).

As for driving environments, EyeSim supports two standard input formats: *world*-files and *maze*-files. The maze format is the simpler of the two, allowing character graphics to construct a driving environment. The characters "_" and "|" represent horizontal and vertical walls. As an example, we can easily create an empty rectangle (Figure 2.9, left).

But we could just as easily specify a maze like the ones that are used for the Micromouse competition [Christiansen 1977][4] that the robot has to navigate through. Figure 2.9, right, shows an example of a competition maze with an

---

[4] D. Christiansen, *Spectral Lines – Announcing the Amazing MicroMouse Maze Contest*, IEEE Spectrum, vol. 14, no. 5, SPEC 77, May 1977, p. 27 (1)

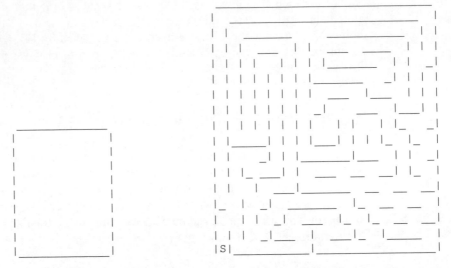

Figure 2.9: Rectangle and maze environments from character graphics file

"*S*" marking the robot's starting place, while the goal is always in the center. More on mazes and how to get out of them will follow in Chapter 9.

2.7 Display and Input Buttons

All of our real robots carry a touchscreen display on top, which is extremely valuable for displaying data (e.g. sensor values and results), entering parameter values, selecting and starting programs, and so on.

Of course, we have the same functionality in EyeSim – in fact, the simulation system runs the same source code as the physical robots. Below are the most important RoBIOS API commands for the display:

```
int LCDPrintf(const char *format, ...)
int LCDSetPrintf(int row, int col, const char *format, ...)
int LCDMenu(char *st1, char *st2, char *st3, char *st4)
int LCDClear(void)
int LCDPixel(int x, int y, COLOR col)
int LCDLine(int x1, int y1, int x2, int y2, COLOR col)
int LCDArea(int x1,int y1, int x2,int y2, COLOR c,int fl)
int LCDImage(BYTE *img)
int LCDImageGray(BYTE *g)
int LCDImageBinary(BYTE *b)
```

Most of these commands should be self-explanatory. They are for writing text onto the screen, writing text into a specific row and column, labeling the menu buttons (soft keys), clearing the display, setting a pixel / line / area of specific color, and displaying a full image onto the screen in either color, grayscale or binary format. RoBIOS API commands for reading out the push-button input (soft keys) are

```
int KEYGet(void)      // Blocking read for key
int KEYRead(void)     // Non-blocking read
int KEYWait(int key)  // Wait until key is pressed
```

This allows us to enter user commands and also to wait for a confirmation key to be pressed. Program 2.7 shows a simple "Hello, world!" program in Python that combines these two features.

Program 2.7: Hello world robot program in Python

```
1   from eye import *
2
3   LCDPrintf("Hello from EyeBot!")
4   LCDMenu("DONE","BYE","EXIT","OUT")
5   KEYWait(ANYKEY)
```

The program writes a text line ("Hello ...") onto the screen, labels each of the four soft keys and then waits for the user to press any of the four buttons. Note that without the last *KEYWait* command the program would immediately terminate, clearing any information that was written onto the display. So this is a good method to make sure a program only terminates when desired.

The equivalent application in C is shown in Program 2.8. It has the compulsory *main* function, which could also be included in the Python program. The actual RoBIOS commands are the same as before.

Program 2.8: Hello world robot program in C

```
1   #include "eyebot.h"
2
3   int main()
4   { LCDPrintf("Hello from EyeBot!");
5     LCDMenu("DONE", "BYE", "EXIT", "OUT");
6     KEYWait(ANYKEY);
7   }
```

Running (and compiling) this program results in the screen output shown in Figure 2.10 (for both Python and C).

2.8 Distance Sensors

So far, we have talked about the basics of controlling a mobile robot, but we are missing some essential sensor inputs. We start with reading the infrared PSD (position sensitive devices) sensors, for which the RoBIOS API is

```
int PSDGet(int psd) // Read distance in mm from sensor
```

There are four predefined sensors: *PSD_FRONT, PSD_LEFT, PSD_RIGHT,* and *PSD_BACK* with matching locations on the robot.

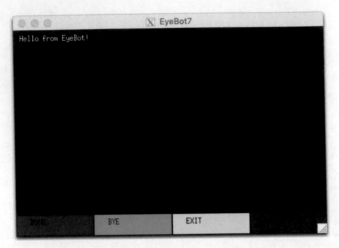

Figure 2.10: Hello world output on (simulated) robot screen

We can incorporate these sensors into our first program for driving straight, but this time stopping the robot before it hits the wall. Program 2.9 shows the code in Python.

Program 2.9: Drive-and-stop program (version 1) in Python

```
1    from eye import *
2
3    while PSDGet(PSD_FRONT) > 200:
4      VWSetSpeed(100,0)
5    VWSetSpeed(0,0)
```

The robot should keep driving while there is at least 200mm clearance in front of it. When this is no longer the case, it will stop by setting speeds to (0,0).

Note that we do not have to repeat the *VWSetSpeed* command inside the loop. We can set it once at the beginning and then have an empty "busy wait loop" instead. This is the correct way of doing it, even though it does not look as nice (Program 2.10).

Program 2.10: Drive-and-stop program (version 2) in Python

```
1    from eye import *
2
3    VWSetSpeed(100,0)
4    while PSDGet(PSD_FRONT) > 200:    # empty wait
5    VWSetSpeed(0,0)
```

The corresponding application in C is shown in Program 2.11. It is not much different from the Python version except for the brackets and semicolons. Here, especially, the semicolon following the *while*-condition is important, as this denotes an empty (wait) statement.

Program 2.11: Drive-and-stop program in C

```
1   #include "eyebot.h"
2
3   int main()
4   { VWSetSpeed(100,0);                    /* drive */
5     while (PSDGet(PSD_FRONT) > 200) ; /* wait  */
6     VWSetSpeed(0,0);                       /* stop  */
7   }
```

If you wanted to save some CPU time, you could insert something like *OSWait(100)* for waiting 100 milliseconds (0.1s):

```
while (PSDGet(PSD_FRONT) > 200) OSWait(100);
```

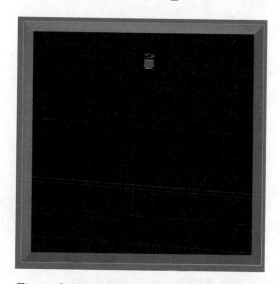

Figure 2.11: Robot driving and stopping in time

In all cases, the robot now avoids a collision and stops 200mm in front of the wall (Figure 2.11).

Writing the PSD sensor value to the screen is a very good idea to help in debugging any robot program. Program 2.12 shows the slightly augmented code. *LCDMenu* and *KEYWait* frame the rest of the code to avoid the program terminating (and erasing the display) when the robot comes to a stop. We use a new variable *dist* for reading and checking the distance to avoid calling the *PSDGet* routine twice in each loop iteration.

The equivalent application in C is shown in Program 2.13. C provides a *do-while* loop, which checks the termination condition at the end of the loop, making coding of this example more elegant.

Program 2.12: Drive-and-stop program displaying distances in Python

```
1    from eye import *
2
3    LCDMenu("","","","END")
4    VWSetSpeed(100,0)
5    dist = 1000
6    while dist > 200:
7      dist = PSDGet(PSD_FRONT)
8      LCDPrintf("%d ", dist)
9    VWSetSpeed(0,0)
10   KEYWait(ANYKEY)
```

Program 2.13: Drive-and-stop program displaying distances in C

```
1    #include "eyebot.h"
2
3    int main()
4    { int dist;
5
6      LCDMenu("", "", "", "END");
7      VWSetSpeed(100,0); /* drive */
8      do
9      { dist = PSDGet(PSD_FRONT);
10       LCDPrintf("%d ", dist);
11     } while (dist > 200);
12     VWSetSpeed(0,0);    /* stop  */
13     KEYWait(ANYKEY);
14   }
```

The screen output now updates the robot's wall distance while it is moving. Once the distance value of 200mm has been reached, the robot's movement (as well as the printing to the LCD) is stopped (see Figure 2.12).

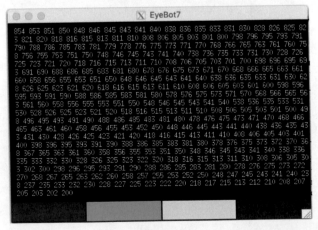

Figure 2.12: Distances printed on robot screen until it stops at 200mm

There are other types of distance sensors, both in simulation as well as on the real robots. Similar to a PSD, but providing much richer data, is a Lidar (Light Detection and Ranging) sensor. It is a rotating laser scanner, which returns several thousand distance points per scan, similar to a large number of PSDs placed in a circle. We will talk more about Lidar sensors in Chapter 4.

Another sensor group is measuring odometry data. These are incremental encoders on the robot's motor shafts, which allow the robot to calculate its position and orientation from combining its left and right encoder values – assuming there is no wheel slip, of course. RoBIOS functions *VWSetPosition* and *VWGetPosition* already translate odometry data into a robot's pose (position and orientation), using also the robot's wheel size, encoder ticks per revolution and wheel distance for the calculation. This data is stored in an HDT (hardware description table) file on a real robot and in a ".robi" definition file on a simulated robot. Feel free to explore these additional sensors.

2.9 Camera

Finally, we introduce a robot's most important sensor, the camera. Each of our real and simulated robots is equipped with a digital camera. In simulation, the camera's position and orientation can be set through the robot's ".robi" definition file and it can be placed on top of a (real or simulated) pan-tilt actuator, which allows us to rotate the camera in two axes during use. The RoBIOS API for reading a camera image is:

```
int CAMInit(int resolution)     // Set camera resolution
int CAMGet(BYTE *buf)           // Read color camera image
int CAMGetGray(BYTE *buf)       // Read gray scale image
```

The camera has to be initialized first using *CAMInit*, which also sets the desired camera resolution. The most common values for this are VGA (640×480), QVGA (quarter-VGA, 320×240) or QQVGA (quarter-quarter-VGA, 160×120). On simulated as well as on real robots, it is often best to start with the low QQVGA resolution to get a proof of concept running, as it requires the least processing time.

In the Python example in Program 2.14, *CAMInit* initializes the camera as QVGA, so each subsequent call to *CAMGet* will return an array of 320×240 color values. Each color value has three bytes, one each for a pixel's red, green and blue (RGB) component.

With a *main* function added in Python, similar to a C program, this requires a few extra lines of code (Program 2.15). The same application in C is shown in Program 2.16. Being more explicit, the C program clearly defines variable *img* as an array of predefined size *QVGA_SIZE*, which is internally specified as 320×240×3 (for three bytes per pixel). Variable *img* can then be used as a parameter for *CAMGet* and *LCDImage*. In C we check the termination condition at the end of the loop, while in Python we can only check it at the start.

While the robot is not moving by itself in this example, you can grab it with the mouse and move it around the driving environment to see the changes on

Program 2.14: Simple camera program in Python

```
1   from eye import *
2
3   LCDMenu("", "", "", "END")
4   CAMInit(QVGA)
5   while (KEYRead() != KEY4):
6       img = CAMGet()
7       LCDImage(img)
```

Program 2.15: Simple camera program with main function in Python

```
1   from eye import *
2
3   def main():
4       LCDMenu("", "", "", "END")
5       CAMInit(QVGA)
6       while (KEYRead() != KEY4):
7           img = CAMGet()
8           LCDImage(img)
9
10  main()
```

Program 2.16: Simple camera program in C

```
1    #include "eyebot.h"
2
3    int main()
4    { BYTE img[QVGA_SIZE];
5      LCDMenu("", "", "", "END");
6      CAMInit(QVGA);
7      do { CAMGet(img);
8           LCDImage(img);
9      } while (KEYRead() != KEY4);
10     return 0;
11   }
```

the display. Placing a few more objects into the scene as shown in Figure 2.13 will add to the fun.

2.10 Robot Communication

Robots can communicate with each other via the Raspberry Pi's built-in WiFi module. We implemented some basic communication commands for sending messages from robot to robot, between a group of robots, or to and from a base station. These will work on the real robots as well as in simulation. The most important communication commands are

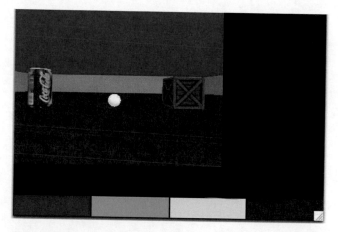

Figure 2.13: Camera program output

```
int RADIOInit(void)                    // Start communication
int RADIOGetID(void)                   // Get own radio ID
int RADIOSend(int id, char* buf)       // Send string to dest.
int RADIOReceive(int *id_no, char* buf, int size) // Rec.
int RADIOStatus(int IDlist[])          // Get rob ID list
```

A simple ping-pong token transmission example for two robots demonstrates the usage of this interface. Robot no. 1 will send message "1" to robot no. 2; after that, each robot returns the received message incremented by one.

Setting up the *SIM* script, we need two robots to work together. We choose two different robot types (S4 and LabBot), but they are both running the same program (Program 2.17).

Program 2.17: SIM script for running two robots

```
1    # robotname  x y phi
2    S4         400  600   0  ping.x
3    LabBot     1000 600 180  ping.x
```

In the application in Program 2.18, we first start the communication interface with *RADIOInit*, and then retrieve the robot's unique ID number in the network. In simulation, robots will always be labelled 1, 2, 3 and so on; however, the ID number for real robots is derived from a robot's IP address, so it will depend on the network.

The robot with ID number 1 is deemed to be the master[5] and immediately sends message "A" to his partner robot no. 2. From then on, the program is identical for both robots. Each robot waits for the next message coming in with *RADIOReceive*, increments the first character in the message (so robot no. 2

---

[5] Note that for the real robots, we cannot guarantee that a robot with this number actually exists, so we need to use the *RADIOStatus* function to find out all robot IDs in the network and make the one with the lowest number the master.

Program 2.18: Radio communication program in C

```
1   #include "eyebot.h"
2   #define MAX 10
3
4   int main ()
5   { int   i, my_id, partner;
6     char buf[MAX];
7
8     RADIOInit();
9     my_id = RADIOGetID();
10    LCDPrintf("my id %d\n", my_id);
11    if (my_id==1)     // master only
12    { partner=2;      // robot 1 --> robot 2
13      RADIOSend(partner, "A");
14    }
15    else partner=1;  // robot 2 --> robot 1
16
17    for (i=0; i<10; i++)
18    { RADIOReceive(&partner, buf, MAX);
19      LCDPrintf("received from %d text %s\n", partner, buf);
20      buf[0]++;       // increment first character of message
21      RADIOSend(partner, buf);
22    }
23    KEYWait(KEY4);   // make sure window does not close
24  }
```

will generate a "*B*") and then sends it back. This will run for 10 times until both programs terminate. The screenshot in Figure 2.14 shows the printout of both robots side by side.

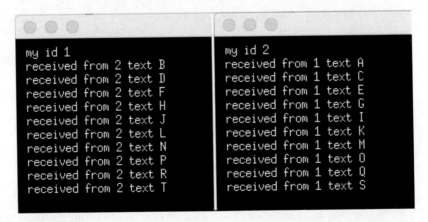

Figure 2.14: Screen outputs for robot 1 (left) and robot 2 (right)

2.11 Multitasking

Running several tasks in parallel makes a lot of sense for robotics applications – even if the processor would have to serialize them. We typically have several different control loops, which have to run at different speeds. For example, one loop that is reading PSD sensors has to run very quickly to avoid a collision, while the time-consuming image processing can run at a slower pace (see Figure 2.15). Although none of the simple examples in the following chapters use multitasking, it is still an essential component for more complex robot programs.

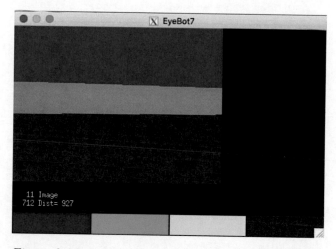

Figure 2.15: Independent iteration counters for camera and PSD sensors

We use the standard *pthreads* (POSIX Threads[6]) package for multitasking, for which a lot of independent literature is available. In Program 2.19, we use a *mutex* (short for *mutual exclusion lock*) to synchronize two threads that are running in parallel.

The main program initializes threads and mutex, then starts two slave threads in parallel (*cam* and *psd*), and finally waits for a button press to terminate the whole program. Each slave thread runs an infinite loop reading sensor data (camera for *cam*, PSD for *psd*) and printing it to the screen, together with a loop counter (*i* and *j*). Access to RoBIOS functions must be framed inside a *mutex_lock* and *mutex_unlock* bracket in order to prevent incorrect results and unpredictable behavior. The mutex lock operation will only let one parallel thread through, while the second one has to wait for the first thread to do the unlock operation. Both threads use *sleep/usleep* to free up processing time.

---

6   POSIX Threads, Wikipedia, https://en.wikipedia.org/wiki/POSIX_Threads

Program 2.19: Multitasking program using pthreads in C

```
1    #include "eyebot.h"
2    pthread_mutex_t rob;;
3
4    void *cam(void *arg)
5    { int i=0;
6      QVGAcol img;
7      while(1)
8      { pthread_mutex_lock(&rob);
9          CAMGet(img);
10         LCDImage(img);
11         LCDSetPrintf(19,0, "%4d Image ", i++);
12       pthread_mutex_unlock(&rob);
13       sleep(1);  // sleep for 1 sec
14     }
15     return NULL;
16   }
17
18   void *psd(void *arg)
19   { int j=0;
20     while(1)
21     { pthread_mutex_lock(&rob);
22         d = PSDGet(PSD_FRONT);
23         LCDSetPrintf(20,0, "%4d Dist=%4d ", j++, d);
24       pthread_mutex_unlock(&rob);
25       usleep(50);  // sleep for 0.1 sec
26     }
27   }
28
29   int main()
30   { pthread_t t1, t2;
31     XInitThreads();
32     pthread_mutex_init(&rob, NULL);
33     CAMInit(QVGA);
34     LCDMenu("", "", "", "END");
35     pthread_create(&t2, NULL, cam, (void *) 1);
36     pthread_create(&t1, NULL, psd, (void *) 2);
37     KEYWait(KEY4);
38     pthread_exit(0); // will terminate program
39   )
```

2.12 Using an IDE

There are a number of excellent IDEs (integrated development environments) available that are free for academic use for Python as well as for C/C++. These environments make program design and debugging a whole lot easier. They allow single-stepping through source code, setting breakpoints and examining variable contents, which are invaluable tools for program development that significantly increase productivity.

Using an IDE

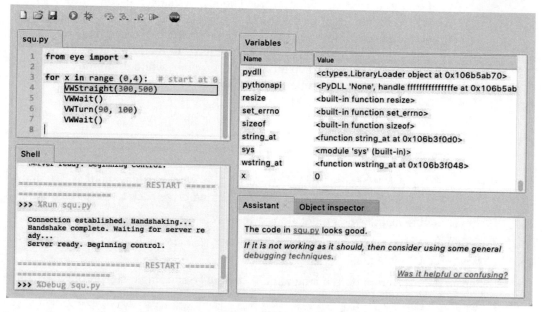

Figure 2.16: Thonny Python IDE

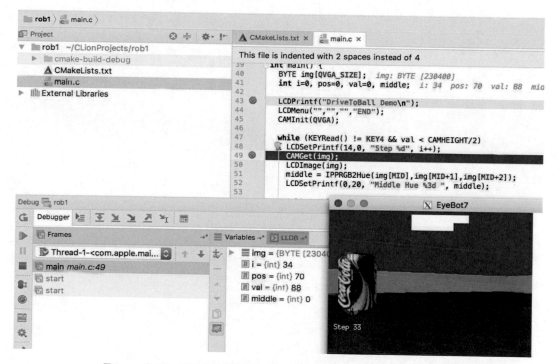

Figure 2.17: CLion C/C++ IDE

For Python, good choices are Thonny[7] (small self-contained package, shown in Figure 2.16) or PyCharm[8] (fully-fledged comprehensive package). Make sure to set the Python interpreter to *Python3* before starting.

For C and C++, a good package is CLion[9]. In the example in Figure 2.17, CLion is used to single-step through a C program while examining changes in variables as well as the robot's LCD output.

2.13 Tasks

- Write a program in Python, C or C++ to drive the robot straight, until it is within 300mm of an obstacle or wall, then let it turn 180° and drive back to its starting point.

- Write a program that drives a robot along a full circle with 1m diameter.

- Extend the ping program using RoBIOS function *RADIOStatus* to make it work for arbitrary robot ID numbers.

- Set up an IDE of your preferred programming language and single-step through a robot program.

---

[7] Thonny – Python IDE for beginners, https://thonny.org
[8] PyCharm – The Python IDE for Professionals and Developers, https://www.jetbrains.com/pycharm/
[9] CLion – A cross-platform IDE for C and C++, https://www.jetbrains.com/clion/

DRIVING ALGORITHMS

3

E ven without any obstacles in its way, driving a robot from point *A* to point *B* can be a challenge. We will first look at an aimless *random drive* before we examine several methods on how to drive to a specific point. Things will get even more complex if we need to arrive at the destination with a specific orientation.

3.1 Random Drive

The first generation of robot vacuum cleaners used a really simple algorithm in an endless loop:

- Drive straight until hitting an obstacle
- Turn a random angle

If you are worried about the cleaning quality of such a behavior, you might be right. But from a mathematical point of view, if given infinite time, this algorithm will cover the complete cleaning area, as long as the robot can physically reach it.

As you can see from Evan Ackerman's long-exposure time photos of cleaning robots with an LED attached (Figure 3.1), there is a lot of unproductive zigzagging going on [Ackerman 2010][1] and [Ackerman 2016][2].

More advanced cleaning robots can drive a much more efficient pattern as shown in Figure 3.2. The robot orientates itself along the major room walls and then drives a regular *lawn mower pattern*.

The program for a random drive (Program 3.1) basically consists of a single *while* loop that runs until the *END*-button (*KEY4*) has been pressed. We wait 100ms (0.1s) in each loop iteration to reduce the compute overhead. An *if*-selection checks whether there is enough space (300mm) to all three sides

---

[1]   E. Ackerman, *Robot Roomba 560 vs. Neato XV-11*, IEEE Spectrum, June 2010, https://spectrum.ieee.org/automaton/robotics/home-robots/irobot-roomba-560-vs-neato-xv11
[2]   E. Ackerman, Review: *Neato BotVac Connected*, IEEE Spectrum, May 2016, https://spectrum.ieee.org/automaton/robotics/home-robots/review-neato-botvac-connected

Figure 3.1: Roomba 880 cleaning pattern [Ackerman 2016]
Photo courtesy of Evan Ackerman / IEEE Spectrum 2016

Figure 3.2: Roomba 980 cleaning pattern [Ackerman 2016]
Photo courtesy of Evan Ackerman / IEEE Spectrum 2016

before continuing to drive forward. If not, the robot backs up a short distance (25mm) and then turns a random angle. Function *random()* produces a number between 0 and 1, so the term

```
180 * (random() - 0.5)
```

produces a value between –90 and +90, which defines the possible range of our random turns. In the next loop iteration, the robot will drive straight again, provided there is enough space along its new direction.

Program 3.1: Random drive program in Python

```
1    from eye import *
2    from random import *
3
4    safe=300
5    LCDMenu("","","","END")
6
7    while(KEYRead() != KEY4):
8        OSWait(100)
9        if(PSDGet(PSD_FRONT)>safe and PSDGet(PSD_LEFT)>safe
10       and PSDGet(PSD_RIGHT)r>safe):
11           VWStraight(100,200)
12       else:
13           VWStraight(-25,50)
14           VWWait()
15           dir=int(180*(random()-0.5))
16           VWTurn(dir,45)
17           VWWait()
```

Program 3.2 uses an extended version of the same algorithm. It prints the numerical values of the PSD sensors onto the display, which requires them to be stored in variables *f*, *l*, and *r*. It also prints a message of the robot's action (always a good idea) whenever it turns, and it continuously reads a camera image and displays it as well. The screenshot in Figure 3.3 shows the robot driving after a couple of straight legs. We used the robot soccer playing field as a background for this task.

The C version of this program closely resembles the Python implementation; it just uses the different syntax. The driving result is exactly the same – see Program 3.3. The camera is initialised as QVGA and its images are displayed along with the PSD readings for front, left and right. A press of KEY4 ("END" soft key) is required to terminate the program.

As the robot will stop for an obstacle of any kind, it will also stop when it encounters another robot. So we can now safely let several robots run in the same programming environment. To do this, we only have to add one extra line for each robot into the *SIM* script. In the script in Program 3.4, we are starting three different robots – two *LabBots* and one *SoccerBot S4*. All of the robots in this example have the same executable program, but you can easily specify different programs by changing the executable filename.

Figure 3.3: Robot executing random drive program

Program 3.2: Random drive with distance sensor output in Python

```
1    from eye import *
2    from random import *
3
4    safe=300
5    LCDMenu("","","","END")
6    CAMInit(QVGA)
7
8    while(KEYRead() != KEY4):
9        OSWait(100)
10       img = CAMGet()
11       LCDImage(img)
12       f=PSDGet(PSD_FRONT)
13       l=PSDGet(PSD_LEFT)
14       r=PSDGet(PSD_RIGHT)
15       LCDSetPrintf(18,0,"PSD L%3d",l,f,r)
16
17       if(l>safe and f>safe and r>safe):
18           VWStraight(100,200)
19       else:
20           VWStraight(-25,50)
21           VWWait()
22           dir = int(180*(random()-0.5))
23           LCDSetPrintf(19,0,"Turn  %d",dir)
24           VWTurn(dir,45)
25           VWWait()
26           LCDSetPrintf(19,0,"     ")
```

Program 3.3: Random drive program in C

```
1    #include "eyebot.h"
2    #define SAFE 300
3
4    int main ()
5    { BYTE img[QVGA_SIZE];
6      int dir, l, f, r;
7
8      LCDMenu("", "", "", "END");
9      CAMInit(QVGA);
10
11     while(KEYRead() != KEY4)
12     { CAMGet(img);     // demo
13       LCDImage(img);   // only
14       l = PSDGet(PSD_LEFT);
15       f = PSDGet(PSD_FRONT);
16       r = PSDGet(PSD_RIGHT);
17       LCDSetPrintf(18,0, "PSD L%3d F%3d R%3d", l, f, r);
18       if (l>SAFE && f>SAFE && r>SAFE)
19         VWStraight(100, 200); // start driving 100mm dist.
20       else
21       { VWStraight(-25, 50); VWWait(); // back up
22         dir = 180 * ((float)rand()/RAND_MAX-0.5);
23         LCDSetPrintf(19,0, "Turn %d", dir);
24         VWTurn(dir, 45); VWWait(); // turn [-90, +90]
25         LCDSetPrintf(19,0, "            ");
26       }
27       OSWait(100);
28     } // while
29     return 0;
30   }
```

Program 3.4: SIM script file for multiple robots in the same environment

```
1    # Environment
2    world $HOME/worlds/small/Soccer1998.wld
3
4    settings VIS TRACE
5
6    # robotname x y phi
7    LabBot  400   400    0 randomdrive.py
8    S4      700   700   45 randomdrive.py
9    LabBot 1000  1000   90 randomdrive.py
```

Adding one extra line per robot is easy, but even this can get tedious if you want to have, let's say, 100 robots for a swarm application. For these applications, there are generic methods for the *SIM* script available, which we will talk about in Chapter 5 on robot swarms. The three random drive robots are shown at various stages of their journey in Figure 3.4.

Figure 3.4: Three robots executing random drives in a shared environment

3.2 Driving to a Target Position

The opposite of *random drive* is driving towards a target or goal. We now introduce a number of different methods that let us drive from A to B, assuming there are no obstacles between these points. Later in Chapter 10, we will look at the more complex scenarios that include obstacles that need to be avoided.

Figure 3.5 shows a number of possible methods for how to get from point A (robot position in the top left) to point B (red dot at bottom right). We can

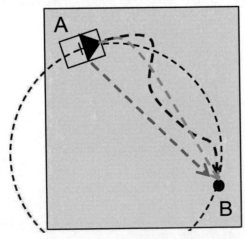

Figure 3.5: Driving methods straight, circle, dog curve and spline

- turn on the spot until we have the correct heading, then drive straight towards the goal (dark green line),
- drive along the arc of the circle that links A to B (blue line),
- constantly incrementally change the robot's heading to home in on the target ("*dog curve*", light green line),
- drive along a calculated cubic spline curve that also allows us to specify the desired robot orientation when arriving at the goal (red line).

We will now look at each of these methods in more detail.

3.3 Turn and Drive Straight

Rotating on the spot followed by a straight-line drive is probably the simplest method for getting from A to B. Although the robot drives the shortest distance, it is probably not completing the task in the shortest possible time, as it executes two separate motions and has to come to a full stop after completing the turn before it can start driving straight.

Program 3.5 shows the algorithm. First, we calculate the angle relative to the goal by using the function *atan2*. Unlike *atan* which takes the quotient dy/dx as a single argument, function *atan2* takes the parameters dy and dx as two separate parameters and can therefore calculate the correct unique angle:

$$goal\_angle = atan2(dy, dx)$$

As the function returns a result in unit *rad*, we have to transfer it into degrees before we can use it for the *VWTurn* function.

For calculating the driving distance we use the Pythagorean formula

$$goal\_distance = \sqrt{(dx^2 + dy^2)}.$$

Program 3.5: Rotate and drive straight in C

```
1   #include "eyebot.h"
2   #define DX 500
3   #define DY 500
4
5   int main()
6   { float angle, dist;
7     // calculate angle and distance angle = atan2(DY,DX);
8     angle = atan2(DY, DX) * 180/M_PI;
9     dist  = sqrt(DX*DX + DY*DY);
10
11    // rotate and drive straight
12    VWTurn(angle, 50);      VWWait();
13    VWStraight(dist, 100);  VWWait();
14  }
```

The actual driving commands then become very simple. We use *VWTurn* with the calculated angle followed by *VWStraight* with the calculated distance value. Note that both drive commands have to be followed by a call to function *VWWait*, which halts the main program execution until the drive command has finished.

Figure 3.6 shows the result of the turn-and-straight method with a marker placed at the desired goal location. The simulated robot does not hit the marker exactly because of its slightly inaccurate turning operation, which is similar to the behavior of a real robot. This problem could best be solved by using sensor input, such as vision or Lidar, to continually update the relative target position (compare with Figure 11.10 in Chapter 11 on robot vision).

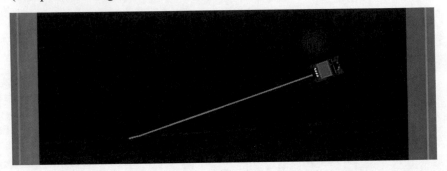

Figure 3.6: Driving to a target position using the turn-and-straight method

3.4 Circle

Instead of rotating and then driving straight, we can calculate the required angular speed from the distance between points A and B, in combination with the angular difference between line AB and the robot's initial heading. We can then issue a single driving command with a constant curvature, forming a circle arc.

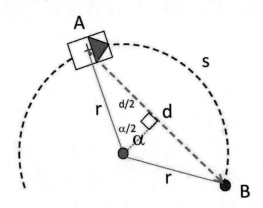

Figure 3.7: Driving arc calculation

As before, we use the function *atan2* to calculate the goal direction and the Pythagorean formula for the direct goal distance d. The total rotation angle α is given by the goal direction minus the robot's initial heading *phi*. As is shown in Figure 3.7, we can form a right-angled triangle using half of the line *d* and half of the angle α. Applying the sine formula for α/2 gives us

$$sin(\alpha/2) = (d/2) / r .$$

Solving for radius *r* results in

$$r = d / (2*sin(\alpha/2))$$

and then we can calculate the desired arc length *s* as

$$s = r * \alpha .$$

Program 3.6: Driving a circle using VWSetSpeed in C

```
1    #include "eyebot.h"
2    #define GOALX 1000
3    #define GOALY  500
4    #define SPEED  300
5
6    int main()
7    { float goal_angle, alpha, d, r, s, omega;
8      int   x,y,phi, dx,dy;
9
10     goal_angle = atan2(GOALY, GOALX); // unit is [rad]
11     VWGetPosition(&x,&y,&phi);         // angle in [deg]
12     alpha = goal_angle - phi*M_PI/180;// relative to rob.
13
14     d = sqrt(GOALX*GOALX + GOALY*GOALY);// segment length
15     r = d / (2*sin(alpha/2));          // radius
16     s = r * alpha;                     // arc length
17
18     omega = (alpha * 180/M_PI) / (s/SPEED); // angle/time
19     VWSetSpeed(SPEED, round(omega));
20
21     do
22     { OSWait(100);
23       VWGetPosition(&x,&y,&phi);
24       dx = GOALX-x;   dy = GOALY-y;
25     } while (sqrt(dx*dx + dy*dy) > 100);
26     VWSetSpeed(0, 0);  // stop robot
27   }
```

We now implement all these formulas in Program 3.6. Using the function *VWSetSpeed* we need to calculate a fixed angular speed ω that is matching the selected constant speed *v*. We do this by dividing the total turn angle α by the driving time, which in turn is the distance *s* divided by the linear speed *v*.

We only issue a single *VWSetSpeed* drive command and then check the goal distance in a loop. When the robot is close enough we stop it.

Although correct in principle, this approach does not give good driving results, as the turn function in both simulated and real robots is not perfect (as well as based on integers instead of floating point numbers). A better perform-ing and much simpler solution is the built-in function *VWDrive* that directly implements the desired driving function along a circle. Program 3.7 lists the simple two-line code and Figure 3.8 shows the execution screenshot.

Program 3.7: Driving a circle using VWDrive in C

```
1   int main()
2   { VWDrive(GOALX, GOALY, SPEED);
3     VWWait();
4   }
```

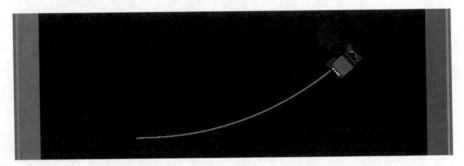

Figure 3.8: Driving along the arc of a circle

3.5 Dog Curve

If a robot maintains a constant speed and initially drives straight from its starting orientation, but then in every step corrects its angle towards the goal, we end up with a continuous movement where the curvature changes in every single iteration step. The resulting path is often called a *dog curve*, suggesting that dogs follow this principle when chasing a target. The algorithm is still quite simple and is shown in Program 3.8.

As before, the goal coordinates are given as a relative offset to the robot's current position (*GOALX, GOALY*). In a *do-while* loop we calculate the robot's current offset to the goal in (*dx, dy*) and then use these values to calculate the distance and angle towards the goal position. The difference between the goal angle and the robot's current heading angle is used in a very simple way for determining the required angular speed ω of the robot:

- If the difference is greater than the threshold of 5° then increment ω
- If the difference is less than the threshold of −5° then decrement ω
- Otherwise set ω to 0

With these values, *VWSetSpeed* can be called with a constant linear speed *v* and the calculated angular speed ω. The loop continues while the robot is more than 100mm away from the goal position, then a stop command is issued and the program terminates. Figure 3.9 shows the resulting *dog curve*.

Program 3.8: Driving a dog curve in C

```
 1    #include "eyebot.h"
 2    #define GOALX 1000
 3    #define GOALY  500
 4
 5    int main()
 6    { float diff_angle, goal_angle, goal_dist;
 7      int    steer=0, x,y,phi,  dx,dy;
 8
 9      do
10      { VWGetPosition(&x,&y,&phi);
11        dx = GOALX-x;   dy = GOALY-y;
12        goal_dist  = sqrt(dx*dx + dy*dy);
13        goal_angle = atan2(dy, dx) * 180/M_PI;
14        diff_angle = goal_angle - phi;
15        if (diff_angle >  5) steer++;
16          else if (diff_angle < -5) steer--;
17            else steer = 0;
18        VWSetSpeed(100, steer/2);
19        OSWait(100);
20      } while (goal_dist > 100);
21      VWSetSpeed(0, 0);  // stop robot
22    }
```

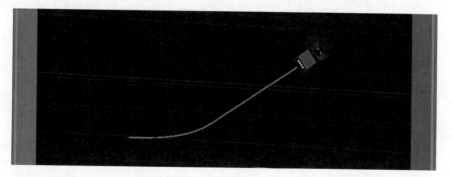

Figure 3.9: Driving along a dog curve

3.6 Splines

Cubic splines are a more complex method for driving from A to B, but they offer a feature that none of the previous methods can. Splines allow us to specify the orientation in the target point B, so the robot will arrive at the specified point with a specified heading. This is quite important for a number of applications; for example, in robot soccer we want the robot to drive to the ball, but it should approach it from an angle where it can kick the ball towards the opponent's goal.

Hermite splines use a parameter u that runs from 0 to 1 over the length of the path to be driven. It uses four blending functions H_1 to H_4 with definitions as follows [Wikipedia 2019][3]:

$$H_1(u) = 2u^3 - 3u^2 + 1$$
$$H_2(u) = -2u^3 + 3u^2$$
$$H_3(u) = u^3 - 2u^2 + u$$
$$H_4(u) = u^3 - u^2$$

The graphical representation of the blending functions is displayed in Figure 3.10. As can be seen, H_1 gradually decreases from one down to zero, while H_2 does the opposite. H_3 and H_4, both start and finish at zero with lower values in opposite directions.

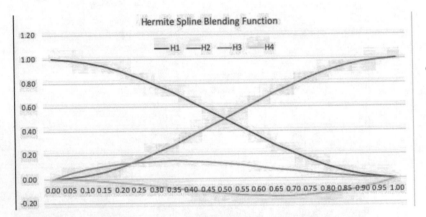

Figure 3.10: Hermite spline blending functions

For the start point a with local pose[4] $[0,0, 0]$ and the goal point b with pose $[x,y, \alpha]$ we set the path length to the direct Euclidean distance multiplied by a scaling factor k:

$$len = k * \sqrt{(x^2 + y^2)}$$

With this, we can initialize start and end points (a_x, a_y and b_x, b_y) as well as their scaled tangent vectors (Da_x, Da_y and Db_x, Db_y). For any point p and angle α, the scaled tangent vector will be

$$Dp_x = len * cos(\alpha) \text{ and}$$
$$Dp_y = len * sin(\alpha) .$$

However, since the local orientation for the robot's starting position a is always 0° (and $cos(0)=1$ and $sin(0)=0$), the start tangent vector becomes simply (len, 0):

start: $a_x = 0$, $a_y = 0$ $Da_x = len$, $Da_y = 0$
goal: $b_x = x$, $b_y = y$ $Db_x = len*cos(\alpha)$, $Db_y = len*sin(\alpha)$

---

3   Wikipedia, *Cubic Hermite spline*, 2019, en.wikipedia.org/wiki/Cubic_Hermite_spline
4   A *pose* combines position and orientation of an object. The pose of a robot moving in 2D space is the translation in x and y and a single rotation angle α.

Next, we iterate the formula with parameter u in the range [0, 1] to receive all intermediate points $s(u)$ of the spline curve:

$$s_x(u) = H_1(u)*a_x + H_2(u)*b_x + H_3(u)*Da_x + H_4(u)*Db_x$$
$$s_y(u) = H_1(u)*a_y + H_2(u)*b_y + H_3(u)*Da_y + H_4(u)*Db_y$$

We can then plot the generated points with a spreadsheet application in Figure 3.11. The higher the scaling factor k, the further the spline curve deviates from the straight line between a and b.

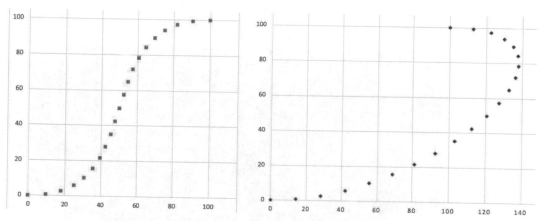

Figure 3.11: Spline points for destinations [100,100, 0°] with scaling factor 1.5 (left) and [100, 100, 180°] with scaling factor 2 (right)

The implementation of the intermediate spline point generation follows directly from the definitions made above. Program 3.9 shows the coding in C.

Program 3.9: Spline point generation in C

```
 1    for (float u = 0.0; u <= 1.0; u += INTERVAL) // [0..1]
 2    { u2 = u*u;   u3 = u2*u;
 3
 4      h1 =   2*u3 - 3*u2 + 1;
 5      h2 = -2*u3 + 3*u2;
 6      h3 =     u3 - 2*u2 + u;
 7      h4 =     u3 -   u2;
 8
 9      sx = ax*h1 + bx*h2 + Dax*h3 + Dbx*h4;
10      sy = ay*h1 + by*h2 + Day*h3 + Dby*h4;
11    }
```

Driving along the generated points is another problem altogether. We steer the robot using the difference between the robot's current heading, read from its localization function *VWGetPosition*, and the desired heading, derived from the line between the previous and the current spline point. Driving the robot can then be done piecewise using *VWCurve* with the short distances between intermediate spline points (see Program 3.10).

Program 3.10: Spline driving function in C

```
1    sphi = round(atan2(sy-lasty, sx-lastx) * 180.0/M_PI);
2    VWGetPosition(&rx,&ry,&rphi);
3    VWCurve(DIST, sphi-rphi, SPEED);
4    VWWait();
5    lastx=sx; lasty=sy;
```

Figure 3.12 shows the final drive along a spline curve for destination pose [1450, 650, 0°].

Figure 3.12: Driving along a spline curve

3.7 Tasks

- Write a *SIM* script to start three different robot programs. Robot-1 should go left–right from one goal to the other and back, robot-2 should drive up and down the middle line and robot-3 should do a random drive, starting in the top-left corner.
  All robots should stop and back up on encountering an obstacle or another robot.

- Change the *SIM* script so that all three robots run the same executable file. Combine the three source programs into one and let a call to function *OSMachineID* decide which part each robot should execute.

- Complete the spline driving program and make it flexible by accepting command line parameters for any destination pose [*x*, *y*, α].

- Improve all A-to-B driving routines in this chapter by replacing fixed coordinates with sensor-based object detection functions in every step of the iteration.

LIDAR SENSORS

4

Lidar stands for "light detection and ranging". A Lidar sensor has one or more rotating laser beams and can generate several thousand distance points per revolution in a fraction of a second. Typical automotive Lidar sensors have 8, 16 or 32 separate beams that allow a much better interpretation of the 3D environment.

Distance data from a Lidar sensor is much simpler to process than from a camera, as it directly provides distance information, whereas image data requires complex computations from stereo or motion sequences in order to extract distances. Lidar sensors are installed on most autonomous research vehicles, including the most successful driverless cars to date – the Waymo fleet (formerly Google X). Unfortunately, Lidar sensors are very expensive. Even a single-beam Lidar for robotics applications costs several thousand dollars, while a multi-beam automotive Lidar can cost up to $100,000.

High-quality Lidar sensors measure the time of flight for each reflected beam and calculate the spatial distance accordingly. As these beams travel with the speed of light, this requires high-performance timing circuits, which explains the high sensor costs. Lidar sensors for measuring only a single point often use a simpler refraction displacement technology, which is cheaper to implement and is often used for electronic distance measurement devices in the building industry.

4.1 Lidar Scans

We are placing a robot in a square driving environment that has one inward-facing corner. By default for this S4 robot type, the Lidar scans a full 360° clockwise, starting from and ending at the *back* of the robot. Figure 4.1 indicates the sensor rotation, while Figure 4.2 visualizes the sensor range data. It generates 360 distance values, so it has a resolution of 1°.

The software for generating the Lidar scan is just a single function call to *LIDARGet*. In our C application (Program 4.1) we plot the distance values on the LCD screen. For each of the 360 values, we draw a blue line of correspond-

© Springer Nature Switzerland AG 2020
T. Bräunl, *Robot Adventures in Python and C*, https://doi.org/10.1007/978-3-030-38897-3_4

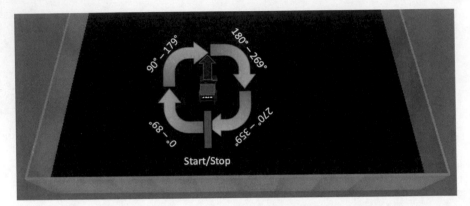

Figure 4.1: Lidar scan range and angles in relation to robot orientation

ing length (scaled down by factor 10) from left to right on the screen, from *x*-position 0 to 359. The *y*-position 250 is at the bottom of the drawing area, leaving some space for printing text plus the input button row.

Program 4.1: Lidar scan and display in C

```
 1   #include "eyebot.h"
 2
 3   int main ()
 4   { int i, scan[360];
 5
 6     do
 7     { LCDClear();
 8       LCDMenu("SCAN", "", "", "END");
 9       LIDARGet(scan);
10       for (i=0; i<360; i++)
11         LCDLine(i,250-scan[i]/10, i,250, BLUE);
12     } while (KEYGet() != KEY4);
13   }
```

To further improve readability of the diagram, we can also add fixed lines for 90°, 180° and 270° in different colors to the diagram (Program 4.2). However, we label them relative to the robot's forward position as –90°, 0° and +90° instead. This extra code gets inserted at the end of the *while*-loop.

Program 4.2: Displaying auxiliary lines and text for Lidar output in C

```
 1   LCDLine(180,0, 180,250, RED);   // straight (0°)
 2   LCDLine( 90,0,  90,250, GREEN); // left    (-90°)
 3   LCDLine(270,0, 270,250, GREEN); // right   (+90°)
 4   LCDSetPrintf(19,0,"      -90        0        +90");
```

Figure 4.2: Lidar scan visualization

In the LCD plot in Figure 4.3, we can now detect the five outward corners as the five local peaks in the diagram at roughly −120°, −30°, +10°, +70° and +110°. The inward facing corner is the sharp local minimum at roughly +30°.

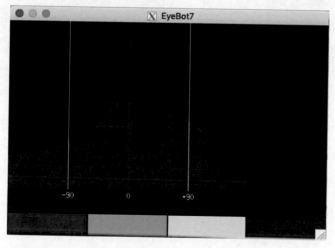

Figure 4.3: Lidar scan plotted on robot's LCD

A similar output can be achieved in Python by Program 4.3. Note that Python loops need to check the condition at the beginning.

Program 4.3: Lidar scan and display in Python

```
1    from eye import *
2
3    LCDMenu("SCAN", "", "", "END")
4    while KEYGet() != KEY4:
5      LCDClear()
6      LCDMenu("SCAN", "", "", "END")
7      scan = LIDARGet()
8      for i in range(90,270):
9         LCDLine(i,250-int(scan[i]/10), i,250, BLUE)
```

4.2 Corners and Obstacles

This section shows a few more Lidar examples. First, in Figure 4.4, there is a robot in the center of a simple square. The Lidar scan shows four uniformly high and evenly distributed peaks – one for each corner.

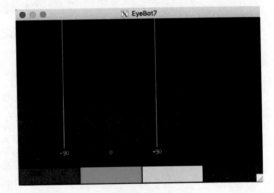

Figure 4.4: Robot placement and Lidar plot for center position

Next, in Figure 4.5, we place the robot into the bottom left corner of the same environment. We still get four local peaks, one for each corner, but here they have different distances (different heights in the diagram), and they are no longer equidistant (left–right) in the diagram as they occur at different angles than before.

If we bring the robot back to the middle of the square and then place a soda can to each side, we will get the scan diagram shown in Figure 4.6. The cans block any information that is radially behind them, which can be clearly seen in the Lidar visualization in Figure 4.6, left. In the Lidar plot on the LCD (Figure 4.6, right), the cans appear as two clearly recognizable cut-outs.

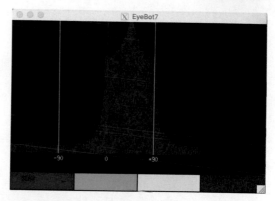

Figure 4.5: Robot placement and Lidar plot for corner position

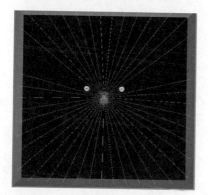
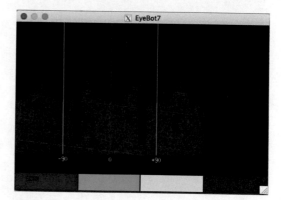

Figure 4.6: Robot visualization with two obstacles and Lidar plot

4.3 Tasks

- Design a simple driving environment world file and store its geometric model on the robot.

- Write a Lidar program that matches the Lidar image with the stored environment image and highlights all possible robot positions (and orientations).

- Let the robot drive around (e.g. wall following or random drive) and from the new Lidar data coming in, eliminate more and more possible positions/orientations until the one correct position/orientation remains.

ROBOT SWARMS

I n the previous chapters we showed how multiple robots can run in the same environment with the same or a different control program. If the number of robots gets very large, specifying them with one line each in the *SIM* script can get lengthy. We therefore introduce a swarm notation that works by using single character placeholders in a maze-format environment file, together with a matching *SIM* script.

5.1 Setting up a Swarm

We start with the maze environment file. The example in Figure 5.1 has 16 placeholders marked by character *a* arranged in a 4×4 grid.

```
 _____
|                          |
|   a    a    a    a       |
|                          |
|   a    a    a    a       |
|                          |
|   a    a    a    a       |
|                          |
|   a    a    a    a       |
|_____|
```

Figure 5.1: Maze environment file for 16 identical robots

In the *SIM* script in Program 5.1, we use this environment file and specify the use of robot type S4 for each of the *a* placeholders. No (x,y)-position placement is necessary here, as the positions are already given in the environment file. If we do not specify an initial orientation either, all robot orientations will be chosen at random. And, as specified by the last parameter, all robots exe-

© Springer Nature Switzerland AG 2020
T. Bräunl, *Robot Adventures in Python and C*, https://doi.org/10.1007/978-3-030-38897-3_5

cute the same program *simple.x*. Figure 5.2 shows the resulting placement of the 16 identical robots

Program 5.1: Repetitive environment SIM script

```
1    # Environment
2    world bots16.maz
3
4    # robotname x y phi
5    S4  a simple.x
```

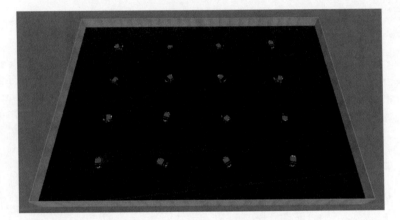

Figure 5.2: Simulation result of 16 identical robots

In the next example (world file in Figure 5.3), we want to use different types of robots, so in the maze environment file we use four different place-holders: *a, b, c* and *d*.

```
 _____
|                               |
|   a     b     c     d         |
|                               |
|   b     c     d     a         |
|                               |
|   c     d     a     b         |
|                               |
|   d     a     b     c         |
|_____|
```

Figure 5.3: Maze environment file for 16 robots of four types

In the matching SIM script in Program 5.2, we first use the "*robot*" construct to load four new, non-standard robots into the environment. After that, we can use their names in the same way as predefined types. So all *a* placeholders become *Cubot*, all *b* placeholders *Cubot-r* and so on. No orientations are given, so the robots will have a random starting orientation. As before, they all share the same executable program. Figure 5.4 shows the result in EyeSim.

Program 5.2: Environment file for four groups of robots

```
 1   # Environment
 2   world bots4x4.maz
 3
 4   # robot definitions
 5   robot ../../robots/Differential/Cubot.robi
 6   robot ../../robots/Differential/Cubot-r.robi
 7   robot ../../robots/Differential/Cubot-b.robi
 8   robot ../../robots/Differential/Cubot-y.robi
 9
10   # robotname placeholder executable
11   Cubot     a   simple.x
12   Cubot-r   b   simple.x
13   Cubot-b   c   simple.x
14   Cubot-y   d   simple.x
```

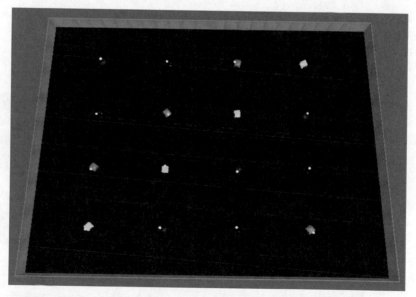

Figure 5.4: Simulation result of 16 robots of four different types

If we want to set up robots with a fixed orientation, we can use the *SIM* script in Program 5.3. It places several S4 and LabBot robots in the same environment. All S4 robots (placeholder *l*) are facing left (orientation 180°), and all LabBot robots (placeholder *r*) are facing right (orientation 0°).

The total number and the individual positions of the robots are determined by the environment file (world format in Figure 5.5). In this case, as you can see, we have placed S4s and LabBots against each other in a friendly match of

Program 5.3: Environment file for five-a-side robot soccer

```
1    # Environment
2    world soccer5-5.maz
3
4    # robotname x y phi
5    S4       1 180 swarm.x
6    Labbot r   0 swarm.x
```

five-a-side soccer. The symbol "*o*" in the middle will be converted to a golf ball, which has the right size for this small-size league robot soccer event. Figure 5.6 shows the resulting scene in EyeSim.

```
    _____
   |                             |   |
   |                             |   |
  _|    r                 1      |_
  |           r        1           |
  | r                o          1 |
  |_         r      1            _|
   |    r                 1      |
   |                             |   |
   |_____|
```

Figure 5.5: Maze environment file for five-a-side soccer

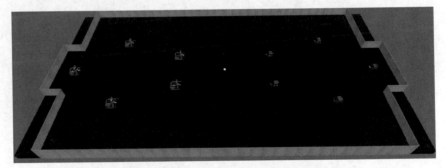

Figure 5.6: Simulation result for five-a-side soccer

5.2 Follow Me

A typical swarm application is to follow a leading robot. We let the lead robot execute its own driving program and then concentrate on the follower robot. The *SIM* script in Program 5.4 defines a LabBot as the leader and an S4 SoccerBot as a follower. For the leader, a one-line program for setting the curve speed is all we need (Program 5.5).

Follow Me

Program 5.4: Leader and follower SIM script

```
1    # # Environment
2    world Field.wld
3
4    # robots
5    Labbot   2000 500 0   leader.py
6    S4        500 500 0   follower.x
```

Program 5.5: Leader program in Python

```
1    from eye import *
2    VWSetSpeed(300, 15)
```

We write the follower program in C for a change (Program 5.6). It uses a Lidar scan instead of the simpler PSD distance sensors to pinpoint the lead robot's exact position. The *LIDARGet* command takes (by default) 360 distance measurements around the robot, which it places into the supplied array. Then a simple loop is used to find the angle with the minimum distance, which is needed for setting the angular speed in the subsequent *VWSetSpeed* command. We use 180° minus the scan angle, as our Lidar works clockwise from the back of the robot, so the array value at position [180] gives the distance straight ahead.

Finally, the *OSWait* statement reduces the speed of the update rate to 10Hz. We need to give the robot a bit of time to execute each drive command before we overwrite it with the next one.

Program 5.6: Follower program in C

```
1    include "eyebot.h"
2
3    int main ()
4    { int i, min_pos, scan[360];
5      while (KEYRead()!=KEY4)
6      { LCDClear();
7        LCDMenu("", "", "", "END");
8        LIDARGet(scan);
9        min_pos = 0;
10       for (i=0; i<360; i++)
11         if (scan[i] < scan[min_pos]) min_pos = i;
12       VWSetSpeed(300, 180-min_pos);
13       OSWait(100); // 0.1 sec
14     }
15  }
```

The screenshots in Figure 5.7 and Figure 5.8 show the successful chase of the follower after the leader robot. Further details on swarm and robot interaction can be found in [Wind, Sawodny, Bräunl 2018][1].

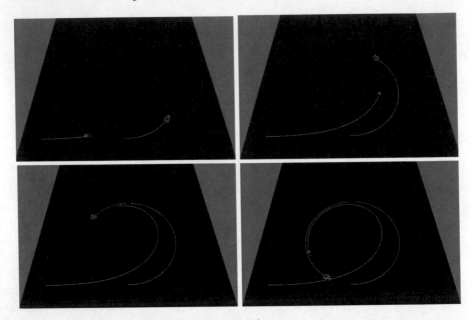

Figure 5.7: Leader-follower steps in simulation

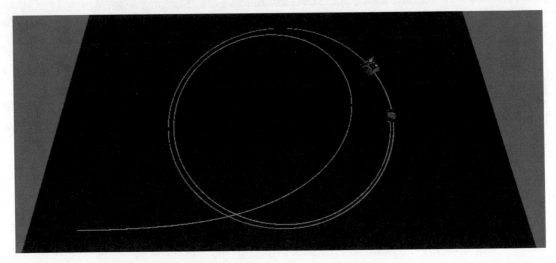

Figure 5.8: Leader-follower scenario in its final stage

[1]   H. Wind, O. Sawodny, T. Bräunl, *Investigation of Formation Control Approaches Considering the Ability of a Mobile Robot*, Intl. Journal of Robotics and Automation, June 2018

5.3 Multiple Followers

Things get more complex if we have multiple robots following the leader. In this case, we have to think how the followers can identify the leader. For example, this could be done by

- radio communication,
  if the leader transmits its current position data to all followers,
- color or shape coding,
  if the leader has a unique color (or shape) that is visible to all followers, or
- high Lidar beam,
  if the leader is higher than all followers and can therefore be detected by an angled or higher placed Lidar sensor.

We chose the latter implementation as shown in the resulting screenshot in Figure 5.9.

Figure 5.9: Lidar positioning above the follower robots

The Lidar sits above the follower robots' height and therefore only detects the leading robot. Please note that on a real robot, the Lidar would need to be slightly angled upwards to eliminate interference with the physical sensors of the other followers. As this would reduce the scanning range to 180° and also limit the maximum detection range of the leader, we will ignore this for now.

In the *SIM* script in Program 5.7, we refer to a new robot type LidarBot, which will have the special Lidar placement discussed before. If we want, we can later add some obstacles into the swarm path, for example

```
can      5000 1400 0
```

Program 5.7: Environment file for self-defined robot type

```
1   # # Environment
2   world field.wld
3   settings VIS
4
5   robot lidarbot.robi
6   ...
```

The Lidar setting is part of a robot's *ROBI* description file, so we have to define our own robot for this application, which we called LidarBot. Descrip-

Program 5.8: Lidar declaration in robot definition file

```
1    # lidar pos relative to robot centre
2    # x y z [mm], rotation x y z [°]
3    # angular range [1, 360], tilt angle [-90, +90], data pts
4    lidar  0 0 100   0 0 0   180 10 180
```

tion file *lidarbot.robi* contains, among many other things, the Lidar specification shown in Program 5.8. Note that the (x,y,z)-displacement is $(0, 0, 100)$, so the Lidar is placed in the center of the robot but 100mm above it. With this the Lidar clears its own robot and other S4-style robots, but will detect the higher handle of the leading LabBot. The scanning range has been set to 180 degrees (centered in front of the robot) and 180 data points, which maintains the $1°$ angular resolution.

Program 5.9: PSD declarations in robot definition file

```
1    # "psd" id, name, pos. to rob-center (right,front,up)  [mm]
2    # R,U,F axis rotations in deg [clockwise is positive]
3    psd 1 PSD_FRONT     0  60 30     0  0    0
4    psd 2 PSD_LEFT     45  60 30     0  0  -90
5    psd 3 PSD_RIGHT   -45  60 30     0  0   90
6    psd 4 PSD_BACK      0 -60 30     0  0 -180
7    psd 5 PSD_FL        0  60 30     0  0  -45
8    psd 6 PSD_FR        0  60 30     0  0   45
9    psd 7 PSD_FFL       0  60 30     0  0  -25
10   psd 8 PSD_FFR       0  60 30     0  0   25
```

We also need PSD distance sensors to avoid collisions with other robots. For this, we define four new PSDs, which point to $25°$ as well as $45°$ diagonally to the front-left and front-right (see Program 5.9 and Figure 5.10).

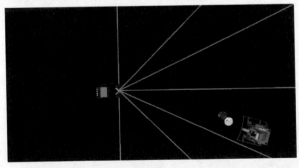

Figure 5.10: Additional PSD sensors for collision avoidance

In the application program, we let the leader just drive straight – again in Python (Program 5.10). We then place five followers behind the leader robot in the SIM script (Program 5.11).

Multiple Followers

Program 5.10: Rotating on the spot in Python

```
1   from eye import *
2   VWSetSpeed(300,0)
```

Program 5.11: Single leader with five followers SIM script

```
1   # # robots
2   Labbot    2500 1200 0   leader-straight.py
3   LIDARBOT  1500 1500 0   follower.x
4   LIDARBOT  1500 1200 0   follower.x
5   LIDARBOT  1500  900 0   follower.x
6   LIDARBOT   500 1400 0   follower.x
7   LIDARBOT   500 1000 0   follower.x
```

The followers' main program in Program 5.12 runs a *while*-loop where it first scans the 180° area in front of the robot using the function *LIDARGet*. It then looks for the minimum value, which will be in the direction of the leader robot's high handle. Variable *min_pos* will then show us the direction of the leader in the range [–90°, +90°]. We also plot the Lidar image onto the screen using *LCDLine*, similar to the single follower scenario before.

Program 5.12: Swarm algorithm for multiple followers in C

```
1    while (KEYRead()!=KEY4)
2    { LCDClear();
3      LIDARGet(scan);
4      min_pos = 0;
5      for (i=0; i<SCANSIZE; i++)
6      { if (scan[i] < scan[min_pos]) min_pos = i;
7        LCDLine(i,250-scan[i]/100, i,250, BLUE);
8      }
9      F  =PSDGet(PSD_FRONT);
10     L  =PSDGet(PSD_LEFT); R  =PSDGet(PSD_RIGHT);
11     FL =PSDGet(PSD_FL);    FR =PSDGet(PSD_FR);
12     FFL=PSDGet(PSD_FFL);   FFR=PSDGet(PSD_FFR);
13     if (F<SAFE) VWSetSpeed(0, -90);
14      else if (L<SAFE || FL<SAFE || FFL<SAFE)
15             VWSetSpeed(150, -20);
16       else if (R<SAFE || FR<SAFE || FFR<SAFE)
17              VWSetSpeed(150, +20);
18        else VWSetSpeed(300, 90 - min_pos);
19      OSWait(200); // 0.2 sec
20     }
```

In the second half of the *while*-loop, we need to check for collisions with other followers, the leader robot or the wall at the end. We could have done this with a second, lower placed or angled Lidar beam, but we decided to use the PSD sensors instead. As discussed before, we added four more PSDs to the standard directions (front, left, right and back) to improve detection and prevention of collisions before they occur (i.e. *FL, FFL, FR, FFR* at ±45° and ±25°). After reading each sensor value, we can determine the new driving direction and speed:

- If the robot comes too close to an obstacle, rotate on the spot.
- If there is an obstacle on the left or right side (90°, 45° or 25°), curve in the opposite direction to avoid it at a reduced speed.
- If all is clear, drive straight towards the leader at full speed.

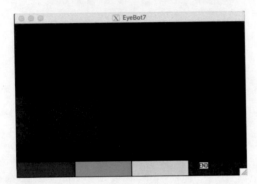

Figure 5.11: Lidar plot of follower robot

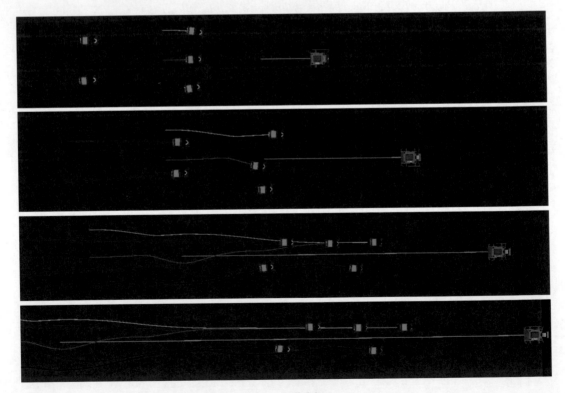

Figure 5.12: Swarm scenario unfolding

A typical Lidar scan looks like Figure 5.11. We can see a distinctive gap in the blue block of 180 scanlines that represents the high handlebar of the leading LabBot, which towers over all the following S4 SoccerBots. This gives us the goal direction, but of course we have to take the PSD data into account to avoid a collision with other followers or the leader.

The screenshots in Figure 5.12 show the motion development of the five robots following their leader. The blue wall to the right side of the driving area prevents the robots from falling off the (virtual) table.

At the end (Figure 5.13), after the leader has been stopped in front of the wall, the follower robots will go off in uncontrolled patterns to avoid a collision while still trying to drive closer.

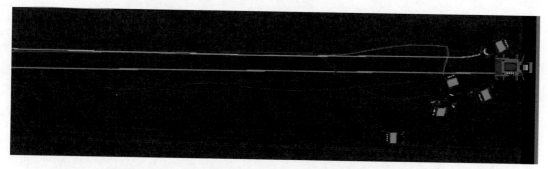

Figure 5.13: Follower robots entangled at the protection wall

5.4 Tasks

- Extend the follower program to drive at a faster speed than the leader. Use Lidar and/ or PSDs to keep a safe distance from it and prevent a collision. Try different driving patterns for the leader robot and see whether the follower can catch it.

- Add obstacles in the driving path and let the leader avoid them. The followers have to avoid all obstacles but still continue their pursuit.

- Use an angled Lidar placement at the top of each follower robot.

6 WALL FOLLOWING

W all following is a good building block for many robotics tasks. It concerns not just the ability of a robot to avoid a collision, but also its detection of the surrounding environment and orientation within it. For this exercise, we use the robot's three PSD sensors, which give distance measurements to the front, left and right. The Lidar sensor with hundreds of distance values may be beneficial for this application, but it is not essential. Considering the high cost of a Lidar, we want to solve this task just using the infrared PSD sensors.

6.1 Wall Following Algorithm

As this algorithm gets a bit more complex, we begin by looking how the wall following algorithm should be executed. As a start, we assume a more or less rectangular driving area with straight walls and right angle corners.

(1) The robot starts in the middle of the field with random orientation to the walls. Not knowing what its orientation towards the environment is, it first drives straight until it encounters a wall and then stops.

(2) The robot now has to turn itself parallel to the wall, leaving it to its left side.

The following two steps are then repeated:

(3) The robot continues driving, "*hugging*" the wall, by continuously updating its driving curve with the help of the left PSD sensor.

(4) When the robot encounters the first corner (detected by the front PSD sensor), it performs a 90° turn.

A perfect run would therefore look like the one in Figure 6.1. Of course, to execute each of these steps properly is far from trivial. Let us go through them one by one.

© Springer Nature Switzerland AG 2020
T. Bräunl, *Robot Adventures in Python and C*, https://doi.org/10.1007/978-3-030-38897-3_6

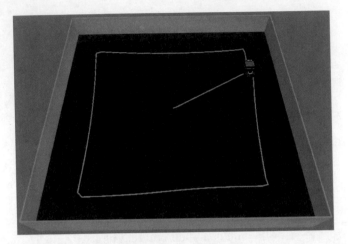

Figure 6.1: Wall following, perfect run

Step 1:   Driving straight can easily be established by the function *VWSet-Speed(x,0)* that specifies a linear speed of *x* and zero rotational speed. Stopping only requires the reading of the front PSD sensor, so implementing this part is simple.

Step 2:   Calculating the correct robot orientation relative to the wall is the key for this step. The subsequent rotation can be easily executed with the functions *VWTurn* and *VWWait*.

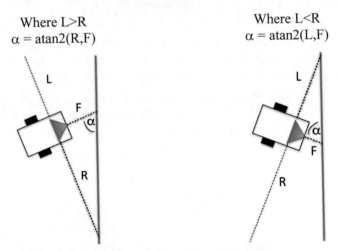

Figure 6.2: Determining the wall angle

As can be seen from the diagrams in Figure 6.2, distance readings from the front and right PSD (left diagram) or the front and left PSD (right diagram) – depending on the approach angle – can be used for determining the rotation angle. As the front and right PSD (or the front and left PSD) are positioned 90° apart, we can use the inverse tangent function to calculate the wall angle α. In software, we will use the *atan2* mathematics function, which takes two arguments (Δy and Δx) instead of the quotient, to make the result angle unique for all input values.

You may have already realized that this rule works in most circumstances but unfortunately not in all. What if the robot is exactly in the middle, between two walls, so $L=R$? Or much worse, what if the robot is so close to a corner that L and R measure distances to two different walls? More work will be required.

Step 3: Driving straight along a wall until the next corner sounds very similar to step 1, although it is not. This would only be the case if step 2 had aligned the robot perfectly with the wall, the wall is exactly straight and the robot is driving perfectly straight – none of which actually occurs in the real world, as there are always some imperfections and noise.

What we need to do instead, is to continuously monitor the robot's distance to the wall, using its left PSD sensor reading to correct the curvature that the robot is driving. Note, however, that the PSD value L may be larger than the robot's actual wall distance d if the robot is not perpendicular to the wall (see Figure 6.3).

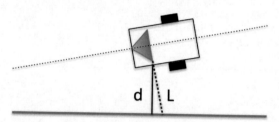

Figure 6.3: Actual wall distance d versus measured distance L

Subsequent measurements of L while driving can be used to determine the robot's angle to the wall and the correct wall distance. This can then be used to correct the robot's driving angle for wall following.

The front PSD can be used as in step 1 to detect the next corner. However, this only works if the robot is not too much out of wall alignment so that the front sensor does not intersect with the wall that is being followed.

Step 4: Turning a fixed 90° angle could be done with *VWTurn*, but we want to be a bit more flexible, as the corner may not be exactly 90° and the robot might not be perfectly aligned with it either. Instead, we suggest that the robot is rotated until there is sufficient clearance (a constant value) in front of *PSD_FRONT*.

Again, this method will work in most cases, but it may not succeed if there are additional obstacles or other robots present. More work may be needed here as well.

6.2 Simplified Wall Following Program

We will now present a simplified program that solves the wall following problem in a simple environment but is far from perfect. Its main program is an endless loop that repeats the two steps *drive* and *turn*, one after the other (Program 6.1).

Program 6.1: Minimal wall following in C

```
1   #include "eyebot.h"
2   #define SAFE   250
3
4   void drive()
5   { do { if (PSDGet(PSD_LEFT)<SAFE) VWSetSpeed(200, -3);
6          else VWSetSpeed(200, +3); // turn right or left
7          OSWait(50);
8        } while (PSDGet(PSD_FRONT) > SAFE);  // next corner
9   }

10
11  void turn()
12  { VWSetSpeed(0, -100);
13    while (PSDGet(PSD_FRONT) < 2*PSDGet(PSD_LEFT))
14      OSWait(50);
15  }

16
17  int main()
18  { while(1)
19    { drive(); turn(); }
20  }
```

So, this means we are using the same "drive straight" function for steps 1 and 3 and the same "turn" function for steps 2 and 4.

The function *drive* does in fact do the wall following. If the *PSD_LEFT* value gets too low, the robot drives a slight right curve (curvature –3), otherwise it drives a slight left curve (+3). So in fact, the robot never drives perfectly straight. An *OSWait* function inside the loop limits the update speed and gives each direction change a little bit of time to have an impact before the next measurement is done. The *PSD_FRONT* measurement is used to terminate the wall-follow routine. This function does not attempt to calculate the robot's real distance from the wall. The result will be a slightly curved wall-follow path.

Using this subroutine for the initial straight drive to the first wall is just to save some lines of code. As can be seen in the drive plot in Figure 6.4, the initial drive leg to the wall is slightly curved; however, it does not really matter either way.

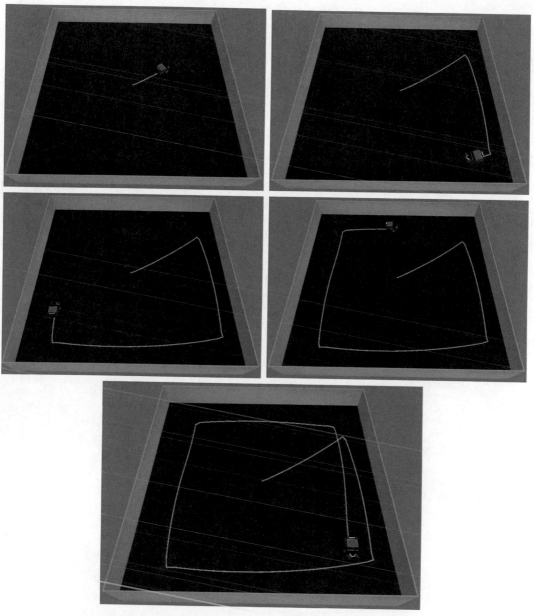

Figure 6.4: Stepwise development of wall following

The *turn* function is even simpler. The robot turns using *VWSetSpeed* until there is sufficient space in front, which we define as twice the space there is to the wall that is being followed (left PSD). Again, calling *OSWait* will limit the cycle speed for the busy-wait loop. Note that none of the functions stop the robot's motion (e.g. by using *VWSetSpeed(0,0)*) when they terminate. They simply assume that the next function will change the robot's speed to whatever is required.

Please note that this algorithm is highly dependent on the environment that the robot drives in and is by no means complete or perfect. It does not behave well in all circumstances and all starting positions or orientations. There are a number of improvements that should be made as outlined in the beginning of this chapter, and this does not even include coping with more complex shaped walls and odd angles. Still, not a bad performance for a program of little more than ten actual lines of code.

6.3 Tasks

- Rewrite the wall following program to solve wall following properly, as outlined in the four steps.
- Create a more complex driving environment, with nooks, angles other than 90°, curved walls, etc. Adapt your program so that it can still do wall following.
- Extend the wall following program to drive a space filling *"lawn mower"* pattern, like the more intelligent vacuum robots shown in Section 3.1 on Random Drive.

ALTERNATIVE DRIVES 7

So far, the vehicle type we have worked with uses a *differential drive* mechanism. It uses two independently driven wheels and there is no need for a steering mechanism as rotation can be achieved by driving one wheel faster than the other. Differential drive is arguably the simplest mechanical drive arrangement; hence, most mobile robots use this system. However, in this chapter we introduce car-like steering, omni-directional wheels and vehicles that can navigate terrain.

7.1 Ackermann Steering

Cars typically have only a single drive motor, whose power is distributed over two wheels (back or front – or all four in all-wheel drive vehicles) by using a mechanical differential. This makes it possible to build a powerful drive system and avoid slip and tire wear when driving curves. However, as driving can only be forward or backward, there is a need for an independent steering mechanism, which will turn both front wheels in the desired driving direction. This so-called *Ackermann steering* is used on all cars, which can be rear-wheel drive, front-wheel drive or all-wheel drive.

Most model cars follow this principle and can therefore be easily modified for autonomous driving. As has been shown in Chapter 1, we can interface a model car directly to a Raspberry Pi controller without the need for any additional hardware and power the controller from a USB power bank. We need two PWM (pulse width modulation) output signals, one for setting the steering angle and one for setting the motor speed. As sensors, we have the Raspberry camera and, if required, a USB Lidar scanner, such as the Hokuyo URG-04LX-UG01 (Figure 7.1). Unfortunately, even small single-line scanning Lidars are not cheap, so they may not be suitable for all projects.

The way a model car is physically connected to the controller will determine which drive commands can be used. If the built-in motor controller is used, the drive system just requires a PWM signal, for which we have the *SERVO*-command in RoBIOS. If the drive motor is controlled from the Eye-

© Springer Nature Switzerland AG 2020
T. Bräunl, *Robot Adventures in Python and C*, https://doi.org/10.1007/978-3-030-38897-3_7

Figure 7.1: Model car with embedded controller, camera and Lidar

Bot I/O board, then the *MOTOR*-command is used (we will assume the latter for now). The steering always requires a PWM signal, which is generated by a *SERVO*-command in RoBIOS.

The subroutine in Program 7.1 combines motor drive and steering commands to a combined function call, assuming the drive motor is linked to motor port 1 and steering is connected to servo output 1 on the I/O board. If two PWM signals are required, they can be linked to servo outputs 1 and 2. In the RoBIOS library, *MOTORDrive* accepts values in the range [−100, +100] for backward or forward driving at variable speeds. Setting the speed to 0 will stop the motor. *SERVOSet* accepts single-byte values in the range [0, 255], with 0 being the servo's far-left position (here for steering), 127 being the neutral middle position (driving straight) and 255 being the far-right position.

Program 7.1: Procedure for drive-and-steer command in C

```
1   void Mdrive(int drive, int steer)
2   { MOTORDrive(1, drive);
3     SERVOSet  (1, steer);
4   }
```

When transforming a model car into a robot without an I/O board, then two of the Raspberry Pi's I/O lines can be used directly for the drive motor and steering. In this case, we recommend using commands from the *wiringPi* library [Wiring Pi 2019][1].

We can then use the subroutine in Program 7.1 to simplify driving commands for the main application (Program 7.2). The *OSWait* statement after each drive command will let the drive command control the vehicle for a few seconds until the next command takes over.

---

[1] Wiring Pi – GPIO Interface library for the Raspberry Pi, 2019, http://wiringpi.com

Program 7.2: Main driving function in C

```
1  int main ()
2  { Mdrive("Forward",      60, 127);  OSWait(4000);
3    Mdrive("Backward",    -60, 127);  OSWait(4000);
4    Mdrive("Left Curve",   60,   0);  OSWait(2000);
5    Mdrive("Right Curve",  60, 255);  OSWait(2000);
6    Mdrive("Stop",          0,   0);
7    return 0;
8  }
```

The screenshot in Figure 7.2 shows the robot's drive from a bird's eye perspective (left) as well as the model car configuration from the side (right). Ackermann drive vehicles can drive curves up to a certain minimum radius but cannot turn on the spot.

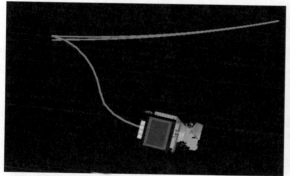 

Figure 7.2: Ackermann driving (left) and side view (right)

7.2 Omni-directional Drives

Differential drive vehicles can drive forward/backward, drive along curves and they can turn on the spot – but they cannot drive sideways. Ackermann steering vehicles cannot even turn on the spot and are quite restricted by their minimum turning curve radius. Therefore, it would be nice to have a vehicle type that is omni-directional, i.e. one that can drive in any given direction.

There are in fact vehicle types that can accomplish this. Most prominent are vehicles with so-called *Mecanum* wheels. These are quite complex mechanical wheel assemblies, which will effectively rotate the wheel's force vector on the driving surface. The interplay of a 3-wheel or 4-wheel Mecanum configuration with as many independent motors allows driving in any given direction, as well as turning on the spot.

The original Mecanum wheel was invented by the Swedish engineer Bengt Ilon and patented in the U.S. (US3876255, submitted 1972 / granted 1975) and Germany (DE2354404, 1973/1974). Our Omni-1 robot (Figure 7.3, left) has been built following this wheel design.

Figure 7.3: Robots Omni-1 (left) and Omni-2 (right)

The surface of each wheel is covered by a number of freely rotating barrel-shaped rollers, which are held by pins from the left and right rim. Only the barrels make contact with the driving surface and they are orientated at –45° to the driving directions on the front-left and rear-right wheels, and at +45° for the front-right and rear-left wheels. These mirrored wheels are physically different from each other and cannot be transformed into each other by rotation.

The ingenious idea behind this wheel design is that the rotational force of the wheel is being split into a force along the rotational axis of the barrel (blue) and the one perpendicular to it (red), as shown in Figure 7.4. The blue force will be eliminated through a small movement of the free roller, so only the red force at ±45° remains.

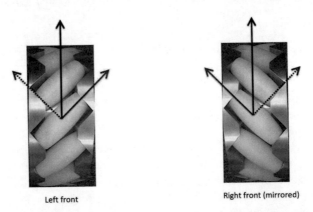

Left front   Right front (mirrored)

Figure 7.4: Forces on front-left and front-right Mecanum wheels

With four Mecanum wheels mounted on a vehicle, we can now observe its overall movement as in Figure 7.5.

1. If all four wheels are moving forward, the vehicle will move forward.

2. If the front-left and rear-right wheels are moving backward, their force vectors will be negated. If the other two wheels are moving forward, then the vehicle's overall movement will be sideways to the left.

3. If the front-right and rear-right wheels are moving backwards and the other two forward, the vehicle will rotate clockwise on the spot.

Figure 7.5: Straight, left, and rotating robot motion with corresponding Mecanum wheel directions

By modifying the individual four wheel speeds, any possible driving angle, combined with any possible angular self-rotation speed, can be achieved. For more details and motion formulas see [Bräunl 2008][2].

Figure 7.6: Mecanum wheel designs for Omni-2 (left) and Omni-1 (right)

[2] T. Bräunl, *Embedded Robotics – Mobile Robot Design and Applications with Embedded Systems*, 3rd Ed., Springer-Verlag, Heidelberg, Berlin, 2008

As the free rollers only protrude a little bit above the rims, this wheel design only works on hard floors, such as concrete or timber, and not on softer surfaces. A subsequent improvement was made by the US Navy, which completely eliminates the rims and instead holds each roller from a stay in the middle. That way, they can navigate softer surfaces as well. Our Omni-2 robot was built following this updated wheel design (see Figure 7.3, right, and Figure 7.6, left).

We will now place a predefined Omni robot in a checkerboard environment so that we can better see its movements (see the SIM script in Program 7.3 and the screenshot in Figure 7.7).

Program 7.3: Omni-robot SIM script

```
1    # Environment
2    world ../../worlds/small/Chess.wld
3
4    # Robot placement
5    Omni 600 600 0 omni-drive.x
```

Program 7.4: Sample Mecanum driving patterns in C

```
1    #include "eyebot.h"
2    void Mdrive(char* txt, int FLeft, int FRight,
3                            int BLeft, int BRight)
4    { LCDPrintf("%s\n", txt);
5      MOTORDrive(1, FLeft);
6      MOTORDrive(2, FRight);
7      MOTORDrive(3, BLeft);
8      MOTORDrive(4, BRight);
9      OSWait(2000);
10   }
11
12   int main ()
13   { Mdrive("Forward",      60, 60, 60, 60);
14     Mdrive("Backward",    -60,-60,-60,-60);
15     Mdrive("Left",        -60, 60, 60,-60);
16     Mdrive("Right",        60,-60,-60, 60);
17     Mdrive("Left45",        0, 60, 60,  0);
18     Mdrive("Right45",      60,  0,  0, 60);
19     Mdrive("Turn Spot L",-60, 60,-60, 60);
20     Mdrive("Turn Spot R", 60,-60, 60,-60);
21     Mdrive("Stop",          0,  0,  0,  0);
22     return 0;
23   }
```

The basic movement code just sets individual wheel speeds for each of the four wheels. The C code is shown in Program 7.4; the Python version is almost identical.

Figure 7.7: Mecanum robot Omni-1 on checkerboard plate

7.3 Driving in Terrain

Many mobile robots can only navigate on a flat surface. However, many practical robotics problems require robots to drive on uneven floors or even in arbitrary 3D terrain. While the mechanical changes on the robot for terrain driving might be relatively simple, the challenge is now on the sensor side as well as on the software side. Path planning in terrain requires a transition from 2D to 3D algorithms.

In the following, we show a couple of examples for driving in 3D terrain in the EyeSim simulation environment. Before we do so, we should explain how an environment file with 3D information can be constructed. The *SIM* script in Program 7.5 is as simple as always. Here, we use a new robot by the name of *Blizzard*, which uses chains instead of wheels. It is a snow truck that we have in real life as a modified model car and we will look at its actual driving program later.

Program 7.5: Terrain definition SIM script

```
1   # Environment
2   world ../../worlds/aquatic/crater.wld
3
4   # Robot
5   robot ../../robots/Chains/Blizzard.robi
6   Blizzard 400 400  0 terrain.x
```

The crater world file follows the *Saphira* world format that we adopted and extended a while ago as one way to input an environment into EyeSim. In this example, the *WLD*-file shown in Program 7.6 specifies a world volume in *x, y* and *z* dimensions, while the relative elevation of each point is taken from the *crater.png* grayscale image, mapping the [0, 255] range of each image pixel onto the specified world height range of [0, 1000]. In addition, we define the water level at a certain height (in this case 200). This will come in handy in

Chapter 8 where we work with submarines. In this example, it just creates a water hazard inside the crater.

Program 7.6: Environment world file for terrain and water level

```
1    terrain 5000 5000 1000 ../heightmap/crater.png
2    water_level 200
```

 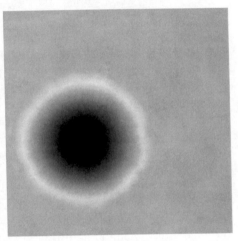

Figure 7.8: Blizzard snow truck (left) and crater graphics file (right)

The PNG image file *crater.png*, which is used to generate the terrain, is just a grayscale image (see Figure 7.8). Each pixel value is being translated into a terrain elevation for the point it represents in 3D space. Further details can be found in the EyeSim documentation. Figure 7.9 shows the resulting screenshot of the *Blizzard* robot driving around the crater landscape.

Program 7.7: SIM script specifying a 3D environment

```
1    # Environment
2    world ../../worlds/aquatic/levels-steel.wld
3
4    # robotname x y phi
5    LabBot 1000 1000  0 terrain.x
```

We would now like to create an environment with ramps that a robot can easily drive up and down, rather than the jagged mountain terrain we had before. Our *SIM* script looks familiar (Program 7.7).

In the world file in Program 7.8 we define a world that creates continuous 3D levels according to the supplied heightmap graphics file *steps.png* (see Figure 7.10, left). A steel texture is used for the graphics effect (Figure 7.10, middle), which will be stretched out over the entire environment.

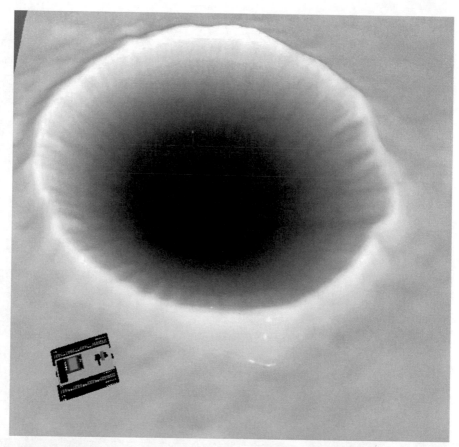

Figure 7.9: Blizzard robot closing in on water-filled crater

Program 7.8: World file for an environment with height levels and texture

```
1    floor_texture ../texture/steel.png
2    terrain 4000 4000 200 ../heightmap/steps.png
```

Any graphics editor can be used to create a heightmap file, even a slide-presentation software will do. In the heightmap file shown in Figure 7.10 (left), black is floor level and white is the highest level. The actual numerical height is specified in the world file (200mm in Program 7.8). So, we have created a vertical ramp up (from dark to bright), a high horizontal bank (white) and a ramp down (bright to dark), all in a square area surrounded by high walls (white).

Changing the texture can be done by just replacing the first line in the world file (Program 7.8), e.g. using the timber texture shown in Figure 7.10, right. Figure 7.11 shows the final 3D environment in a steel texture with a LabBot trying to find its way around the ramp. The inset top right shows the view from

Figure 7.10: Heightmap profile (left) and textures for steel and wood (right)

the robot's on-board camera – the same environment with a *woodgrain finish* is shown in the inset bottom right.

Figure 7.11: Robot environment with display for terrain example

7.4 Tasks

- Write a program to drive an Ackermann steering robot from position (0,0) to specified (x,y) goal coordinates. This can be done as follows:
  - First rotate, then drive
  - Drive along a circle
  - Drive along a "dog curve"
  - Drive along a Hermite spline curve
- Extend the program so that the robot arrives at the correct position (x,y) and also with a specified orientation φ.
- Write a program that drives the Omni robot in a square without turning. Take advantage of the sideways motions.
- Write a software interface that calculates individual wheel speeds for any given driving angle.
- Write a program to let the Omni robot drive in a straight line while continuously rotating about itself.
- Write a program that drives the tracked robot to the highest point in a mountain range. The robot should always re-orient itself towards the maximum gradient. Use PSD or Lidar sensors and create a moderate terrain steepness so that the robot can cope.
- Write a program that can navigate a robot over the ramp shown in this chapter. Add additional PSD sensors pointing down for detecting a cliff – and prevent the robot from falling off.

Boats and Subs

Autonomous underwater vehicles (AUVs) and autonomous boats are an important sector of robotics research with huge commercial potential. We have built a number of autonomous submarines over the years and we have recently put together an autonomous solar boat.

Figure 8.1: Mako AUV

8.1 Mechanical AUV and Boat Structure

Our first AUV is Mako [Bräunl et al. 2004][1], see Figure 8.1, which has four thrusters: two horizontal thrusters for driving forward/backward/sideways in differential drive mode and two vertical thrusters (front and back) for diving. The AUV is naturally buoyant, so it will surface if the two diving motors are stopped.

---

[1] T. Bräunl, A. Boeing, L. Gonzales, A. Koestler, M. Nguyen, J. Petitt, *The Autonomous Underwater Vehicle Initiative - Project Mako*, 2004 IEEE Conference on Robotics, Automation, and Mechatronics (IEEE-RAM), Dec. 2004, Singapore, pp. 446-451 (6)

T. Bräunl, *Robot Adventures in Python and C*, https://doi.org/10.1007/978-3-030-38897-3_8

Our autonomous solar boat uses a raft design, see Figure 8.2, where all electronics and back-up batteries are placed in waterproof tubes supporting a 100W solar panel. Two thruster motors let us drive and steer the boat without the need for a rudder.

Figure 8.2: Solar raft

8.2 Specifying an Underwater Structure

All hydrodynamics equations for simulating AUVs and boats have been included in EyeSim and we have already seen the required environment settings in Section 7.3 on terrain. Specifying a heightmap will structure the ocean floor (or river bed) and specifying the water level will set the overall depth. You can create a mix of an ocean and islands by choosing a terrain level higher than the water level, similar to the water-filled crater we constructed before.

We select the Mako AUV and place it into a pool world (Program 8.1). The pool world file is defined in Program 8.2.

Program 8.1: AUV underwater SIM script

```
1   # Environment
2   world ../../worlds/aquatic/pool.wld
3
4   # Load custom robot
5   robot ../../robots/Submarines/mako.robi
6
7   # Robot position (x, y, phi) and executable
8   Mako 12500 5000 0 mako-dive.x
```

Program 8.2: Pool world file

```
1   floor_texture ../texture/rough-blue.jpg
2
3   terrain 25000 50000 2000 ../heightmap/olympic-pool.png
4   water_level 1900
```

The heightmap (Figure 8.3, left) is a very basic graphics file with only two colors: black (elevation 0) over the whole pool area with a white wall (elevation 255, which here equals 2m) surrounding it. The pool floor texture *olympic-pool.png* (Figure 8.3, right) is a structured water-colored texture that indicates realistic ripples.

Figure 8.3: Pool heightmap and texture

The terrain parameters in the world file (Program 8.2) specify the dimensions in *x* and *y*, as well as the maximum terrain height, which in this case is the wall around the pool (2000mm). The water level is set to 1900mm in the following line of the world file, so it is 100mm below the surrounding pool wall. Figure 8.4 shows the resulting scenario of the Mako submarine swimming in the pool with its local camera view shown as an inset.

Figure 8.4: Mako in pool environment with controller screen

8.3 Submarine Diving

The short example in Program 8.3 only uses Mako's two dive motors: the two side-mounted differential drive thrusters for forward/backward/turn movements are not being used.

Via the menu buttons *KEY1, KEY2* and *KEY3* the user can activate diving, stop diving or breach the submarine. This will actuate both dive motors together in the diving direction, stop them or reverse them. Pushing button *KEY4* will terminate the program.

Program 8.3: Mako submarine diving in C

```
1    #include "eyebot.h"
2
3    #define LEFT    1      // Thruster IDs
4    #define FRONT   2
5    #define RIGHT   3
6    #define BACK    4
7    #define PSD_DOWN 6    // new PSD direction
8
9    void dive(int speed)
10   { MOTORDrive(FRONT, speed);
11     MOTORDrive(BACK,  speed);
12   }

13
14   int main()
15   { BYTE img[QVGA_SIZE];
16     char key;
17
18     LCDMenu("DIVE", "STOP", "UP", "END");
19     CAMInit(QVGA);
20     do { LCDSetPrintf(19,0, "Dist to Ground:%6d\n",
21                             PSDGet(PSD_DOWN));
22          CAMGet(img);
23          LCDImage(img);
24
25          switch(key=KEYRead())
26          { case KEY1: dive(-100);  break;
27            case KEY2: dive(  0);  break;
28            case KEY3: dive(+100);  break;
29          }
30       } while (key != KEY4);
31     return 0;
32   }
```

8.4 Submarine Movement

Only a few lines of code need to be changed to test Mako's movement instead of its diving. Function *drive* in Program 8.4 now takes two parameters – the speeds of the left and right motors. In the main program, we associate the key button activation with driving forward and rotating left or right. To accomplish these movements, we simply specify the left and right motor speeds. In this example, we only use the speed values for full forward, full backward and stop (+100, −100, 0); however, the whole value range can be used for fine-tuning AUV movements.

Program 8.4: Mako driving forward, left and right in C

```
 1   void drive(int l_speed, int r_speed)
 2   { MOTORDrive(LEFT,  l_speed);
 3     MOTORDrive(RIGHT, r_speed);
 4   }

 5
 6   int main()
 7   { LCDMenu("FORWARD", "LEFT", "RIGHT", "END");
 8       ...
 9           switch(key=KEYRead())
10           { case KEY1: drive( 100,  100);  break;
11             case KEY2: drive(-100,  100);  break;
12             case KEY3: drive( 100, -100);  break;
13           }
14       ...
15
```

8.5 Tasks

- Write an AUV program that performs wall following along the sides of a pool.
- Write an AUV program that scans the whole pool area (or ocean floor of a given range) and automatically generates a depth map.
- Change the camera orientation to face down, and then write an AUV program that searches the pool or ocean floor for a specific object, e.g. using color coding.

MAZES

9

Exploring a maze can be great fun – be it walking through a full-size maze or solving it as a puzzle. Solving a maze problem in simulation before trying a run with a real robot is highly recommended, as it helps to greatly reduce software development and debugging time. If the difference between simulated and real robot behavior, the so-called *reality gap*, is sufficiently small, then the transition between simulation and reality will be quite smooth.

Mazes are also a great skill tester for robots. One of the very first competitions for mobile robots was the *Amazing MicroMouse Maze Contest* (see Figure 9.1), which was first proposed in May 1977 in IEEE Spectrum and – after some trial events – held its first final in 1979 in New York City [Allan 1979][1].

Figure 9.1: Competition mazes for London 1986 and Chicago 1986

---

[1] R. Allan, *The amazing micromice: see how they won*, IEEE Spectrum, Sept. 1979, vol. 16, no. 9, pp. 62-65 (4)

T. Bräunl, *Robot Adventures in Python and C*, https://doi.org/10.1007/978-3-030-38897-3_9

9 Mazes

9.1 Micromouse

The Micromouse contest has been a robotics benchmark for generations of engineering and computing students. Although it dates back to 1977, there are still competitions being held today.

The task sounds simple: place the robot in the start square in the bottom left corner of the maze and let it find the central goal square. Whichever drives from start to goal the quickest wins. Of course, the robot must not be told the shortest path, so it requires previous runs to explore the maze and calculate the best route. Each robot gets a total of 10 minutes for the Micromouse competition, and whenever it returns to the start square a new run timer is started – only its shortest run counts.[2]

The full maze is made out of 16×16 cells, which are each 18cm × 18cm in size. All wall segments are 5cm high and 1.2cm thick.

Teams have used all types of approaches to win this competition. The first robots were purely electromechanical without a microprocessor brain. Their technique was dubbed *"wall hugging"* because they were always following the left-most wall, which is a standard way to safely escape any planar maze. Although they did not even attempt to calculate the fastest path, they were faster than the more sophisticated intelligent robots in the 1970s. A rule change that placed the goal in the middle of the maze, without any connecting wall to the rest of it, eliminated this approach.

After that there was an evolution of sensor technology. From sonar sensors to infrared distance sensors, from Lidar to vision. Even using "outrigger-style" sensor placement above the walls was allowed, which helped to detect walls more reliably and more accurately than other methods. Finally, improvements in the drive system and wheel traction were required to make the robots faster. If you watch any recent Micromouse competition live or online, you will be amazed by their speed [YouTube 2017][3].

9.2 Wall Following

As previously mentioned, always following the left-most wall (or always following the right-most wall – but not changing tactics in between) will let you find the exit of a planar maze (which has to be on the outside of a 2D maze), or it will bring you back to the starting point if the entry and exit are at the same location.

We have a simple method of specifying maze environments in the EyeSim simulator by allowing character-graphics environment files. The example in Figure 9.2 is a small maze, which uses character *S* as a placeholder for the robot's starting position so it can be automatically centered in the start square.

---

2   RoboGames, Maze Solving / Micromouse Rules, 2019, robogames.net/rules/maze.php
3   2017 All Japan classic micromouse contest 1st prize, www.youtube.com/watch?v=LAYdXIREK2I

Using an uppercase character such as the *S* in the example assumes that there will also be a wall below the character, which is something we cannot draw using character graphics.

```
| | |___   |
| |___     |
|   _  | |_|
| | | |    |
|S|_____|_|
```

Figure 9.2: Maze input as character graphics

This environment (stored as text file *small.maz*) can then be loaded into the simulator via the *SIM* script in Program 9.1. Placeholder *S* marks the robot's starting position and an orientation of 90° is chosen to let it face upwards (orientation 0° is along the *x*-axis to the right).

Program 9.1: Maze environment SIM script

```
1    # Environment
2    world small.maz
3
4    # Robot description file using "S" as start position
5    S4 S 90 maze_left.x
```

The environment and robot are shown in the screenshot in Figure 9.3. The robot is already engaging its standard three PSD sensors (front, left, right), which are shown as green beams in the visualization.

Figure 9.3: Maze input as character graphics

The full algorithm (Program 9.2) is fairly simply as it has only about ten lines of code, not counting comments and blank lines. In the main *while*-loop, we check for possible walls to the front, left and right by reading out the corresponding PSD sensors as shown in the diagram on the left. These three answers will just be true or false (1 or 0).

The second step is conducting rotations. If the left side is free, we need to make a left turn. Function *VWTurn* will execute a rotation about the given angle and then stop. We need to use *VWWait* following *VWTurn* in order to halt the execution of the program until this motion command has been completed. Otherwise our program would quickly go into the next loop iteration and the next motion command would wipe out the current one. That means the robot would not even start to move.

Program 9.2: Left-hand following in C

```
1    #include "eyebot.h"
2    #define THRES 400
3
4    int main ()
5    { int Ffree,Lfree,Rfree;
6      LCDMenu("","","","END");
7      do
8      { // 1. Check walls
9        Ffree = PSDGet(PSD_FRONT) > THRES;
10       Lfree = PSDGet(PSD_LEFT)  > THRES;
11       Rfree = PSDGet(PSD_RIGHT) > THRES;
12
13       // 2. Rotations
14       if (Lfree)          { VWTurn(+90, 45); VWWait(); }
15        else if (Ffree)    { }
16          else if (Rfree) { VWTurn(-90, 45); VWWait(); }
17            else          { VWTurn(180, 45); VWWait(); }
18
19       // 3. Driving straight 1 square
20       VWStraight(360, 180); VWWait();
21     } while (KEYRead() != KEY4);
22    }
```

In the *else*-case (when there is a wall to the left), we check whether the front is clear. If so, we should go straight, so no rotations are required – hence the empty brackets. The second *else* means that there are walls to the left *and* to the front, so we check whether we can go right. If yes, we turn right (the same sequence of *VWTurn*, only in the opposite direction, followed by *VWWait*).

The final *else*-case means that there must be walls to the left, front and right, so our robot is caught in a dead end. It therefore has to turn 180° before it can move again.

The third step after all these rotations is to move forward by exactly one square. Function *VWStraight* followed by *VWWait* does exactly that. Linear and angular speed are set in a way that the robot completes driving of one square as well as turning ±90° in about two seconds. Also note that we are driving 36cm instead of 18cm as per the Micromouse rules. Our real robots are too large for this maze size, so we just doubled the dimensions.

Figure 9.4: Left-hand following method

Logically, this program is very clear and it should work perfectly, as shown in the diagram in Figure 9.4. Going left wherever possible until it reaches an exit or the start again should work — only, it does not!

Figure 9.5: Left-hand following method

If you look at the screenshot in Figure 9.5, you will get an idea of what happened. The simulated robot (very much like a real one) does not drive perfectly straight nor turn perfectly 90°. Its small driving errors accumulate and it does not take long until the robot collides with a wall. Its spinning wheels make it lose its internal orientation and there will be no recovery from this incident.

9.3 Robustness and Control

The key phrase we are looking for is called *robustness*. We need a robot program that not only works for an unrealistically perfect simulated world, but also for a real robot with its small deviations in actuators and sensors. Our previous program issued driving commands but never checked their results. We need to expand this program to make sure the robot stays centered between two walls when driving and that slightly incorrect turns will be compensated for in subsequent driving commands.

The only part we need to change is the routine for driving straight (step 3). Instead of the code line containing *VWStraight* followed by *VWWait*, we need an extended program as outlined below. In this approach, we are implementing two ideas:

1. **Drive the correct distance**
   a. Limit the driving distance
      In each iteration, we calculate the driving distance using Pythagoras' formula:
      $$drivedist = \sqrt{((x_2{-}x_1)^2 + (y_2{-}y_1)^2)} \ .$$
      We only continue driving while *drivedist* < *MSIZE* (the size of one square in the maze).
   b. Check for front wall
      If the robot has driven a bit too far and the front wall is getting too close, it will also terminate this "driving straight" operation. The minimum front distance should be half the square size (*MSIZE/2*) minus half the robot size, around 50mm. So we only continue to drive if freespace *F* > *MSIZE*/2 − 50 .

2. **Keep the correct wall distance**
   In every iteration, we measure the left and right PSD distance values and we have to distinguish several different cases:
   a. There is an immediate wall to the left and the right
      → Keep the robot exactly in the middle between the two walls.
   b. There is only a wall on the left and a gap to the right
      → Only rely on the left wall distance and the known size of the square to keep the desired robot distance from the wall.
   c. There is only a wall on the right and a gap to the left
      → Similarly, only rely on the right PSD value.
   d. There are gaps to the left and to the right
      → With no immediate wall for orientation, just keep driving straight.

Program 9.3 uses *VWGetPosition* before the loop and at the end of each iteration to calculate the distance driven so far. If the left PSD distance value *L* is in the range [100, 180] at a square size of 360mm, then there must be a wall to the left. If *L* is outside this range, we assume there is a gap; the same applies to distance *R* for the right side.

Program 9.3: Controlled left-hand following in C

```
 1  VWGetPosition(&x1,&y1,&phi1);
 2  do
 3  { L=PSDGet(PSD_LEFT); F=PSDGet(PSD_FRONT);
 4    R=PSDGet(PSD_RIGHT);
 5    if (100<L && L<180  && 100<R && R<180) // check space
 6      VWSetSpeed(SPEED, L-R);    // drive difference curve
 7    else if (100<L && L<180)     // space check LEFT
 8      VWSetSpeed(SPEED, L-DIST);// drive left if left>DIST
 9    else if (100<R && R<180)     // space check RIGHT
10      VWSetSpeed(SPEED, DIST-R); // drive left if DIST>right
11    else                         // no walls for orientation
12      VWSetSpeed(SPEED, 0);     // just drive straight
13    VWGetPosition(&x2,&y2,&phi2);
14    drivedist = sqrt(sqr(x2-x1)+sqr(y2-y1));
15  } while (drivedist<MSIZE && F>MSIZE/2-50); // stop in time
```

The four nested *if-else* selections test for the four cases: both walls available, only left, only right or no wall on either side. In each case, we use the function *VWSetSpeed* for driving the robot. Its two parameters are linear speed (we just use a constant) and angular speed, which we calculate as the difference between the left and right wall distance (if we have both walls) or the difference between the desired wall distance and the actual measured distance (if there is only one wall available).

If there is a wall to both sides, we want the left distance to be equal to the right so that our curve value is $L-R$. This will be zero if the robot is perfectly in the middle. If $L>R$ then $L-R$ will be positive, so the robot will be turning left. And vice versa for $R>L$. Note that this method of steering actually constitutes a proportional control or P-control [Bräunl 2008][4]. The larger the error, the larger the control (steering) value. One can use a proportionality factor, but just using 1 (or omitting it) works perfectly in this case.

If there is only a wall to the left (second case), we use the same function, but we use $L-DIST$ as the desired control value (*DIST* is the desired wall distance). Similarly, $R-DIST$ is used if there is only a wall to the right (third case).

And finally (fourth case), if there are no immediate walls to either side, we continue to drive straight.

As can be seen from the screenshots of traversing the maze in Figure 9.6, the robot does quite a good job, despite not always turning exactly 90°. Some corrections have to be done, which result in a wiggly motion. This can be improved with better control algorithms, such as PID-control (proportional integral differential control), and by using additional sensors, for example, reading the calculated (x,y) position from the wheel encoders via the function *VWGetPosition* to find the robot's displacement. We currently use this only to terminate driving when the next square cell has been reached.

---

[4] T. Bräunl, *Embedded Robotics – Mobile Robot Design and Applications with Embedded Systems*, 3rd Ed., Springer-Verlag, Heidelberg, Berlin, 2008

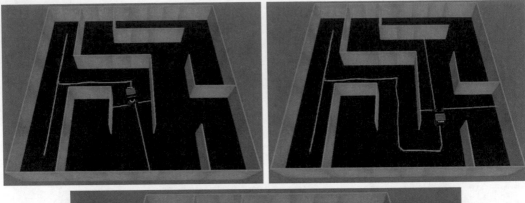

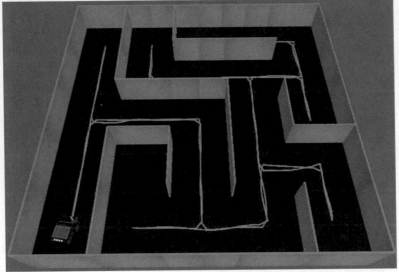

Figure 9.6: Maze-driving with P-controller

9.4 Maze Driving with Lidar

Having only three data points (PSD for front, left and right) is obviously a large restriction. Using a Lidar instead will give us a much richer data volume of several hundred to several thousand distance points. The algorithm for driving with a Lidar is as follows in an endless loop:

1. Conduct a Lidar scan around the left side of the robot, angles [45°, 135°].

2. Find the smallest scan value and corresponding angle. This defines the angle of the left wall location α; see diagram Figure 9.7, left.

3. Use *VWCurve* to drive parallel to the wall.

4. If the robot gets close to the wall in front (scan[180] < THRES) then stop and turn right by 90° using *VWTurn* followed by *VWWait*.

For calculating the correct driving angle, we can look at the diagram in Figure 9.7, right. The robot scans a total area of 90° centered around its left side, which is the angular range [45°, 135°] from the robot's forward perspective. We can then determine the angle and length of the shortest beam (angle α and length s in Figure 9.7). Since the shortest beam will hit the wall at a right angle (90°), we have $\alpha + \beta = 90°$. Ideally, α should be 90° so that β equals 0°, which means driving parallel to the wall. If angle α is less than 90°, then the robot is steering into the wall and needs to veer right; if the angle is greater than 90°, then the robot is steering away from the wall and needs to veer left.

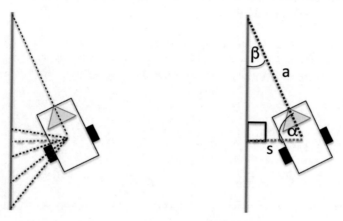

Figure 9.7: Lidar measures left and front of robot (left) and angle calculation from Lidar distances (right)

We need to control two variables here:
- the robot's angle from the wall α, and
- the robot's distance from the wall s.

We have the following sensor values:
- the shortest distance to the left wall s, and
- the front collision distance a (only if $\alpha < 90°$)

For driving the robot, we use the *VWCurve* function. Using a constant distance and linear speed, we only need to control the angular speed. We use a combined *proportional controller* or *P-controller* for the two variables [Bräunl 2008][5]. In a P-controller, a larger error term will result in a larger output value – using a proportionality factor k.

- The first term controls the robot's orientation:
  $$k_1 \cdot (\alpha - 90°)$$
  If the angle is correct at 90°, then the term is zero. A smaller angle (robot steering into the wall) results in a negative value, which will steer the robot to the right. If the angle is larger than 90° (robot drifting away), a positive result will steer it to the left.

[5] T. Bräunl, *Embedded Robotics – Mobile Robot Design and Applications with Embedded Systems*, Springer-Verlag, Heidelberg, Berlin, 2008

- The second term controls the robot's wall distance:

$$k_2 \cdot (s - 250\text{mm})$$

  If the robot is exactly at the desired wall distance of 250mm, then this term is zero. The further the robot is away from the ideal distance, the larger this term gets, which will steer the robot to the left. If the robot is too close to the wall, the term becomes negative and the robot steers to the right, away from the wall.

In practice, we found by experiment good values to be $k_1 = 5$ and $k_2 = 0.5$. Of course there are also formal methods for finding optimal control parameters [Aström, Hägglund 1995][6].

Program 9.4: Lidar maze exploration in C

```
1    while (KEYRead() != KEY4)
2    { LIDARGet(scan);
3      if(scan[180] < 300)   // check for front collision
4        VWTurn(-90, 360); VWWait();   // turn right 90°
5      findMin(scan, 45,135, &angle, &s);   // check left
6      printf("min angle %d val %d\n", angle, s);
7      a = 180-angle;
8      VWCurve(50, k1*(a-90) + k2*(s-250), SPEED);
9    }
```

The code for the Lidar scanning and driving routine is shown in Program 9.4. Note that the Lidar angles count clockwise from the back (back = 0°, front = 180°), so $\alpha = 180° - angle$.

We also plot the detected Lidar points at their correct position to the EyeBot LCD, generating a global environment map, as shown in Figure 9.8. The algorithm design and program implementation was done by Joel Frewin at the UWA Robotics & Automation Lab, and adapted by the author.

The robot successfully traverses the complete maze and at the same time generates its Lidar-based internal representation, shown to the right at each stage in Figure 9.8. These look quite accurate until about three quarters of the way through the maze exploration phase. Unfortunately, the angular dead-reckoning error accumulates over time and leads to incorrect mapping (see the final screenshot of the sequence in Figure 9.8). The same phenomenon happens with real robots and is one of the largest challenges in localization and mapping. The most popular method for solving this problem is a statistical method called Simultaneous Localization and Mapping, or SLAM for short [Durrant-Whyte, Bailey 2006][7], [Bailey, Durrant-Whyte 2006][8].

---

[6] K. Aström, T. Hägglund, *PID Controllers: Theory, Design, and Tuning*, 2nd Ed., Instrument Society of America, Research Triangle Park, NC, 1995

[7] H. Durrant-Whyte, T. Bailey, *Simultaneous Localisation and Mapping (SLAM): Part I*, IEEE Robotics & Automation Magazine, vol. 13, no. 2, June 2006, pp. 99–110

[8] T. Bailey, H. Durrant-Whyte, *Simultaneous Localisation and Mapping (SLAM): Part II*, IEEE Robotics & Automation Magazine, vol. 13, no. 3, Sep. 2006, pp. 108–117

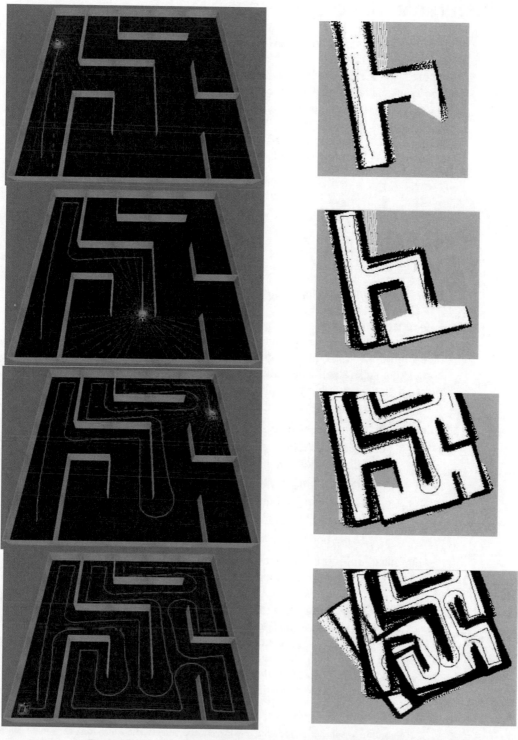

Figure 9.8: Lidar maze drive (left) and internal representation (right)

9.5 Recursive Maze Exploration

Applying the "left-wall following algorithm" cannot solve a Micromouse maze that has the goal is in the center without a connecting wall to the starting square. This means, we have to use a more sophisticated method to explore the *whole* maze and not just a part of it. Since we know that the complete maze is made out of 16×16 square cells, we can create an internal data structure and mark every cell that is visited. That way, we can tell whether we have seen all the cells. Note that depending on the maze construction, some cells may be blocked off completely by walls so they can never be reached.

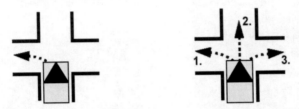

Figure 9.9: Left-hand following (left) vs. recursive exploration method (right)

Arguably the easiest way to fully explore the maze is using a recursive algorithm[9]. So instead of going always left, if possible, at any new square cell (see Figure 9.9, left), all free directions have to be explored. So, if the robot comes to a square cell where the left, front and right are open (see Figure 9.9, right), then the robot has to explore each of these directions, one after the other (but not necessarily in the order left, front, right).

The function *explore* uses recursion to search the maze.

- For directions (left, front, right),
  if the direction is free and has not been visited previously:
  - Drive one square to this free neighbor square.
  - Mark the new square as visited.
  - Recursively call function *explore* from the new position.
  - Drive one square back to the previous position and orientation.

The maze in Figure 9.10 shows the recursive path of the robot. At the first two square cells, there is no choice in exploration, e.g. the second cell only has the front and back walls open, so the robot just passes through. However, the third cell is a *branching point* marked by a red circle, which has two openings besides the entry point – one in front and one to the right. Here, we are not making a choice, instead we need to explore both directions: first we let the robot drive straight (which turns out to be a dead end), and then we let it come back to the square with the red dot and afterwards explore the other direction to the (original) right. We use the data structures shown in Program 9.5.

---

[9] An algorithm or subroutine is called *recursive* if the body of the subroutine calls itself (with different parameters or changed global variables).

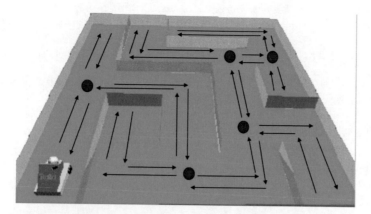

Figure 9.10: Recursive exploration method

Program 9.5: Global variable declarations for maze in C

```
1   int mark[MAZESIZE][MAZESIZE];         // 1=visited
2   int wall[MAZESIZE+1][MAZESIZE+1][2];  // 1= wall, 0=free
3   int map [MAZESIZE][MAZESIZE];         // distance to goal
4   int nmap[MAZESIZE][MAZESIZE];         // copy
5   int path[MAZESIZE*MAZESIZE];          // shortest path
```

The *main* program has a number of tasks:

1. Initialize the variables
2. Call the recursive function *explore*
3. Let the user select a goal square
4. Calculate the shortest path from start square to goal
5. Drive to the goal by the shortest route
6. Drive back to the start

Program 9.6 shows the first part of the code for the exploration function. This is to mark the new square in array *mark*, then read all three PSD sensors. Any walls found are entered into the internal data structure with the function *maze_entry* and function *check_mark* then checks whether the new data have completed all four sides of a yet incomplete square. If so, the robot knows everything there is about this square and also marks it as visited. That way the robot will not have to drive into this square at a later stage and the exploration process can be shortened.

In the second part of *explore* (Program 9.7), the function checks whether any or all of the three directions, front, left and right, are accessible and the square behind it is still unexplored. If so, the function *go_to* will drive the robot in this direction. There, it will call *explore* recursively, after which *go_to* will bring the robot back to the current position. This will guarantee that the robot will explore all possible open directions, one after the other.

Program 9.6: Recursive maze exploration function (part 1) in C

```
 1   void explore()
 2   { int front_open, left_open, right_open, old_dir;
 3
 4     mark[rob_y][rob_x] = 1;     /* mark current square */
 5     left_open  = PSDGet(PSD_LEFT) > THRES;
 6     front_open = PSDGet(PSD_FRONT) > THRES;
 7     right_open = PSDGet(PSD_RIGHT) > THRES;
 8     maze_entry(rob_x,rob_y,rob_dir,          front_open);
 9     maze_entry(rob_x,rob_y,(rob_dir+1)%4, left_open);
10     maze_entry(rob_x,rob_y,(rob_dir+3)%4, right_open);
11     check_mark();
12     old_dir = rob_dir;
13     ...
```

Program 9.7: Recursive maze exploration function (part 2) in C

```
 1     ...
 2     if (front_open  && unmarked(rob_y,rob_x,old_dir))
 3       { go_to(old_dir);     // go 1 forward
 4         explore();          // recursive call
 5         go_to(old_dir+2);   // go 1 back
 6       }
 7
 8     if (left_open && unmarked(rob_y,rob_x,old_dir+1))
 9       { go_to(old_dir+1);   // go 1 left
10         explore();          // recursive call
11         go_to(old_dir-1);   // go 1 right, (-1 = +3)
12       }
13     if (right_open && unmarked(rob_y,rob_x,old_dir-1))
14       { go_to(old_dir-1);   // go 1 right, (-1 = +3)
15         explore();          // recursive call
16         go_to(old_dir+1);   // go 1 left
17       }
18   } // end explore
```

Driving function *go_to* has the desired direction as a parameter, which makes this function a lot more useful. The direction is given as [0, 1, 2, 3] meaning [North, West, South, East].

The function calculates the difference between the desired angle and the robot's current steering angle. If this difference is not equal to zero, the robot will execute the required turn as a multiple of 90° rotations. Subsequently, the robot drives one square forward using the previously described driving method with a P-controller.

Finally, the subroutine in Program 9.8 updates the robot's position and orientation (*rob_x, rob_y, rob_dir*). Instead of using a complex formula, we call our subroutines *xneighbor* and *yneighbor*, which use a switch statement for the four cases to update each coordinate.

Program 9.8: Turning and driving one square cell in C

```
1    void go_to(int dir)
2    { int turn;
3      static int cur_x, cur_y, cur_p;
4
5      dir = (dir+4) % 4;              // keep direction in [0, 3]
6      turn = dir - rob_dir;
7      if (turn)
8      { VWTurn(turn*90, ASPEED); VWWait(); }
9
10     Controlled_Straight(DIST, SPEED; // P-contr. straight
11     rob_dir = dir;
12     rob_x    = xneighbor(rob_x,rob_dir);
13     rob_y    = yneighbor(rob_y,rob_dir);
14   }
```

To avoid 270° turns in one direction, the additional statements shown in Program 9.9 are inserted to reduce it to a 90° turn in the opposite direction.

Program 9.9: Avoiding 270° turns in C

```
1    if (turn == +3) turn = -1;  // turn shorter angle
2    if (turn == -3) turn = +1;
```

The function *explore* will let the robot traverse the maze and at the same time reproduce it in its internal data structure. An example maze structure is shown in Figure 9.11.

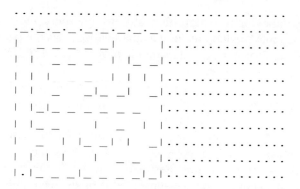

Figure 9.11: Completed internal maze representation

9.6 Flood-Fill

If we are given the goal coordinates by the user (or they are predefined as in the *Micromouse* competition), we need to figure out how to get from S (start) to G (goal). An excellent method for this is a *flood-fill* algorithm for counting the distance steps of every maze square from the start. You can imagine this as pouring water into the starting square of the maze and seeing how long it takes for each square cell to get wet. In the example in Figure 9.12, it will take zero steps for the starting square, one step for the cell above it and two steps for the next, but then we have two cells with three steps, the next one up and the one to the right. They both get token "3" and we continue as shown.

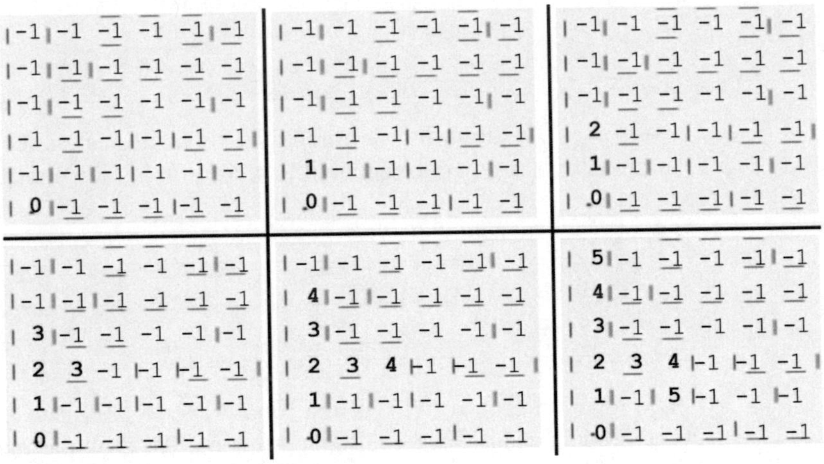

```
|-1|-1 -1 -1 -1|-1    |-1|-1 -1 -1 -1|-1    |-1|-1 -1 -1 -1|-1
|-1|-1|-1 -1 -1 -1    |-1|-1|-1 -1 -1 -1    |-1|-1|-1 -1 -1 -1
|-1|-1 -1 -1 -1|-1    |-1|-1 -1 -1 -1|-1    |-1|-1 -1 -1 -1|-1
|-1 -1 -1|-1|-1 -1    |-1 -1 -1|-1|-1 -1    | 2 -1 -1|-1|-1 -1
|-1|-1|-1|-1 -1|-1    | 1|-1|-1|-1 -1|-1    | 1|-1|-1|-1 -1|-1
| 0|-1 -1 -1|-1 -1    | 0|-1 -1 -1|-1 -1    | 0|-1 -1 -1|-1 -1

|-1|-1 -1 -1 -1|-1    |-1|-1 -1 -1 -1|-1    | 5|-1 -1 -1 -1|-1
|-1|-1|-1 -1 -1 -1    | 4|-1|-1 -1 -1 -1    | 4|-1|-1 -1 -1 -1
| 3|-1 -1 -1 -1|-1    | 3|-1 -1 -1 -1|-1    | 3|-1 -1 -1 -1|-1
| 2 3 -1|-1|-1 -1    | 2 3 4|-1|-1 -1    | 2 3 4|-1|-1 -1
| 1|-1|-1|-1 -1|-1    | 1|-1|-1|-1 -1|-1    | 1|-1|5|-1 -1|-1
| 0|-1 -1 -1|-1 -1    | 0|-1 -1 -1|-1 -1    | 0|-1 -1 -1|-1 -1
```

Figure 9.12: Steps of flood-fill algorithm

The core of the flood-fill algorithm is shown in Program 9.10. The outer *do*-loop calls the inner *for*-loop as long as the goal cell has not been reached and the loop counter does not exceed the number of cells ($MAZESIZE^2$). After this, the distance map is complete.

The inner *for*-loop runs over all square cells in the maze. For each one, it checks the four directions (North, West, South, East). If there is an unmarked accessible neighboring cell (−1), it will mark it with the neighbor's distance value plus one. After the completion of the inner loop, we copy back the distance array to avoid marking more distant cells in the same step.

Eventually, we reach the goal destination, which in this example is the top right cell that happens to have a distance value of 40 (see Figure 9.13).

So, by now we know two things:
- the goal position is reachable from the start, and
- the shortest path from start to the goal has 40 steps.

However, we do not know yet what the actual path is. We will solve this in the next and final step.

Program 9.10: Flood-fill algorithm in C

```
 1     iter=0;
 2
 3     do
 4     { iter++;
 5       for (i=0; i<MAZESIZE; i++) for (j=0; j<MAZESIZE; j++)
 6       { if (map[i][j] == -1)
 7         { if (i>0)
 8             if (!wall[i][j][0]   && map[i-1][j] != -1)
 9               nmap[i][j] = map[i-1][j] + 1;
10           if (i<MAZESIZE-1)
11             if (!wall[i+1][j][0] && map[i+1][j] != -1)
12               nmap[i][j] = map[i+1][j] + 1;
13           if (j>0)
14             if (!wall[i][j][1]   && map[i][j-1] != -1)
15               nmap[i][j] = map[i][j-1] + 1;
16           if (j<MAZESIZE-1)
17             if (!wall[i][j+1][1] && map[i][j+1] != -1)
18               nmap[i][j] = map[i][j+1] + 1;
19         }
20       }
21
22       for (i=0; i<MAZESIZE; i++) for (j=0; j<MAZESIZE; j++)
23         map[i][j] = nmap[i][j];  // copy back
24     } while ( map[goal_y][goal_x] == -1  &&
25                 iter < (MAZESIZE*MAZESIZE) );
```

```
      -1 -1 -1 -1  -1 -1 -1 -1  -1 -1 -1 -1  -1 -1 -1 -1
      -1 -1 -1 -1  -1 -1 -1 -1  -1 -1 -1 -1  -1 -1 -1 -1
Goal  | 8  9 10 11  12 13|38 39 (40) -1 -1 -1  -1 -1 -1 -1
      | 7 28 29 30  31 32|37|40  -1| -1 -1 -1  -1 -1 -1 -1
      | 6 27|36 35  34 33|36|21| 22| -1 -1 -1  -1 -1 -1 -1
      | 5 26 25 24  25|34 35|20  21| -1 -1 -1  -1 -1 -1 -1
      | 4 27|24 23  22 21 20 19  18| -1 -1 -1  -1 -1 -1 -1
      | 3 |12 11 10  11|14 15 16| 17| -1 -1 -1  -1 -1 -1 -1
      | 2  3  4 | 9 12 13|14|15  16| -1 -1 -1  -1 -1 -1 -1
      | 1 | 8 | 5 | 8  9|12 13 14  15| -1 -1 -1  -1 -1 -1 -1
Start (0) 7  6  7 |10 11 12 13| 16| -1 -1 -1  -1 -1 -1 -1
```

Figure 9.13: Steps of flood-fill algorithm

9.7 Shortest Path

Trying to find the path from the start to the goal step-by-step will not work, as there are multiple possible directions at every branching point and we do not know which one to take. For example, if we are going from the start $0 \rightarrow 1 \rightarrow 2$, we then have a 3 to the top and a 3 to the right. We are lost.

Working backwards is the key to finding the shortest path (Figure 9.14). If we start from the goal at distance value 40, then look for a neighboring cell without a wall in between that has value 39, then 38 and so on, we will eventually come back to the starting point at 0. If there was a point where we had a choice (e.g. assume there were two square cells with a value of 37 neighboring the 38), then there would be two different shortest paths. In this case it does not matter which one we choose.

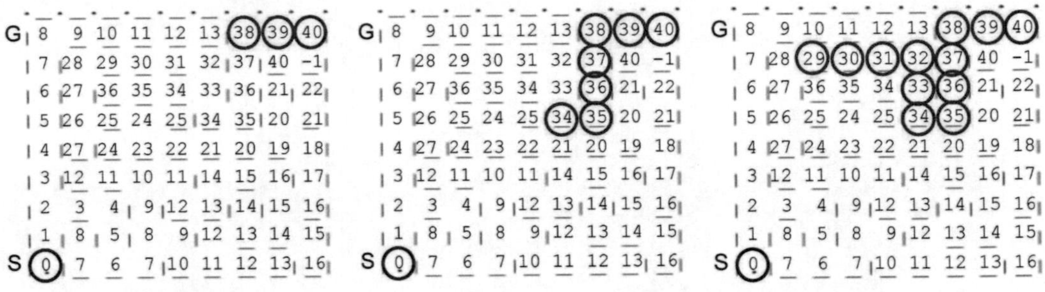

Figure 9.14: Steps of back-to-front pathfinding algorithm

Program 9.11: Shortest path algorithm in C

```
1   void build_path(int i, int j, int len)
2   { int k;
3
4     for (k = len-1; k>=0; k--)
5     { if (i>0 && !wall[i][j][0] && map[i-1][j] == k)
6         { i--;  path[k] = 0; /* north */}
7       else
8         if (i<MAZESIZE-1 && !wall[i+1][j][0]
9             && map[i+1][j]==k)
10          { i++;  path[k] = 2; /* south */}
11        else
12          if (j>0  && !wall[i][j][1] && map[i][j-1] == k)
13            { j--; path[k] = 3; /* east */}
14          else
15            if (j<MAZESIZE-1  && !wall[i][j+1][1]
16                && map[i][j+1] == k)
17              { j++; path[k] = 1; /* west */}
18              else LCDPrintf("ERROR");
19    }
20  }
```

The function for calculating the shortest path is shown in Program 9.11. We use a countdown loop over the known length of the shortest path (40 in this example) and in every step we look for a connected (wall-free) neighboring cell with the current value of k.

Since we have now stored the complete path from the goal back to the start, it will be very easy to just read the path in reverse order and drive the robot along the shortest path from start to finish (Program 9.12). If desired, we can also extend this function to drive the return path back to the start, once the goal has been reached.

Program 9.12: Path recreation algorithm in C

```
1   void drive_path(int len)
2   { int i;
3     for (i=0; i<len; i++)  go_to(path[i]);
4   }
```

After the robot has finished the maze exploration (see Figure 9.6, bottom), the user can select the desired goal position and view the internal maze representation, the distances from the flood-fill algorithm, the calculated shortest path and the map of explored cells marked with an *x* (see Figure 9.15. top left to bottom right). In this example we chose the goal position (4,4), the top right maze square. Nodes with distances −1 were either not required for finding the goal or were not reachable and therefore retain their initialization value. Figure 9.16 finally shows the robot driving along the shortest route to the goal.

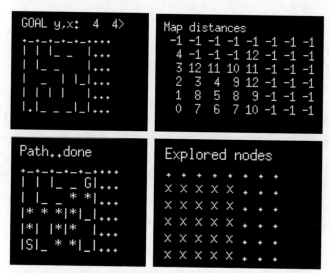

Figure 9.15: Internal representation, flood-fill, shortest path and nodes

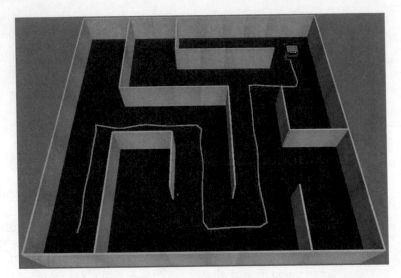

Figure 9.16: Robot driving to the goal along the shortest route

9.8 Tasks

- Combine the controlled driving of the follow-left program with recursive maze exploration so that the robot will find the shortest path.

- Further improve driving control by using a PID sensor as well as incorporating orientation control with position control. This should result in much smoother paths.

- Create a smaller sized robot that can drive diagonally at 45° through subsequent left/right 90° turns and optimize its driving algorithm to take advantage of this.

NAVIGATION

10

I n mobile robot navigation, we distinguish between navigation in known environments, where we are given a map, and navigation in unknown environments [Bräunl 2008][1]. In this chapter, we want to explore algorithms for each of these scenarios.

10.1 Navigation in Unknown Environments

The algorithm we want to look at is called *DistBug* and was developed by [Kamon, Rivlin 1997][2]. It is part of the larger family of *Bug* algorithms, which have the nice proven property that they will find a path if one exists, or they will report that no path exists (after performing a search). The not-so-nice property is that this is a mathematical proof, so it only works for robots of zero size (which we can fix with an offset) and with zero driving and sensing error (which we cannot fix). This means that these algorithms have some robustness issues and may not work perfectly in the real world, but they are definitely worth exploring [Ng, Bräunl 2007][3].

As there is no map in this scenario, the robot will start with local (x,y)-coordinates of $(0,0)$ and an orientation of $0°$. The (x,y)-goal coordinates are given as a relative offset to the robot's starting position, e.g. $(1000,1000)$.

In the following, we present a simplified version of the *DistBug* algorithm:

1. Drive straight towards the goal.

2. If the goal is reached → *finish with success!*

3. If an obstacle is encountered, remember this position (*hit* position), then start wall following around the obstacle (always keeping it on

---

[1] T. Bräunl, *Embedded Robotics – Mobile Robot Design and Applications with Embedded Systems*, 3rd Ed., Springer-Verlag, Heidelberg, Berlin, 2008

[2] I. Kamon, E. Rivlin, *Sensory-Based Motion Planning with Global Proofs*, IEEE Transactions on Robotics and Automation, vol. 13, no. 6, Dec. 1997, pp. 814–822 (9)

[3] J. Ng, T. Bräunl, *Performance Comparison of Bug Navigation Algorithms*, Journal of Intelligent and Robotic Systems, Springer-Verlag, no. 50, 2007, pp. 73-84 (12)

the right side) while constantly calculating the minimum distance to the goal. Continue this operation until either condition holds:

- If there is a clear path to the goal or the robot can advance a minimum distance closer to the goal → call this the *leave* position and go to step 1.
- If the robot is back at the *hit* position → *finish with failure!* (no possible path).

We have already worked with the Lidar sensor, which we need here to measure obstacle distances, and we have done wall following, which is an essential component of this algorithm. In the following section, we will combine them and present the *DistBug* implementation in several parts.

10.2 DistBug Algorithm

The following *DistBug* implementation is by Joel Frewin (UWA), with modifications by the author. In the *SIM* script in Program 10.1, we start the robot at (300,300) and put a colored marker for the goal position on the ground at (4500,4500). The latter is for the viewer only – the robot gets its goal coordinates as an offset from its starting position.

Program 10.1: DistBug environment SIM script

```
1    # World File
2    world obstacles.wld
3
4    settings VIS TRACE
5
6    # Robots
7    S4 300 300 0 distbug.x
8
9    # Objects position x y, color R G B
10   marker 4500 4500  0 255 0
```

We need to include the goal coordinates relative to the robot's starting position (4500–300, 4500–300) in our program, as the robot has no other way of knowing where the goal is (Program 10.2).

Program 10.2: Defining relative goal position in C

```
1    #define GOALX (4500-300)    // marker minus start offset
2    #define GOALY (4500-300)    // (relative goal coord.)
```

The first thing we do in the control loop is to collect all data. We update the Lidar scan all around the robot, get the latest position and orientation data from the wheel encoders via *VWGetPosition* and then calculate the angle towards the goal. For this, we use the math library function *atan2* in Program 10.3, which is a variation of the arcus tangent (*arctan*) function, using the two parameters *dy* and *dx* instead of the quotient *dy/dy*. Function *atan2* returns a unique angle in the range $[0, 2\pi]$, which we convert back to degrees $[0°, 360°]$. We then only need to build the difference between the goal angle and the robot's heading angle to find the goal heading from the robot's current pose (position and orientation).

Program 10.3: Main loop reading Lidar data and calculating goal heading in C

```
1    while (1)
2    { LIDARGet(dists);   // Read distances from Lidar
3      VWGetPosition(&x, &y, &phi);
4      LCDSetPrintf(0,0,"POS x=%5d y=%5d phi=%4d ", x,y,phi);
5      theta = atan2(GOALY-y, GOALX-x) * 180.0/M_PI;
6      if (theta > 180.0) theta -= 360.0;
7      diff = round(theta-phi);
8      LCDSetPrintf(1,0,"GOAL %5d %5d %6.2f diff=%4d ",
9                      GOALX, GOALY, theta, diff);
```

Checking whether the robot has reached the goal is easy. We allow a deviation of 50mm in *x* and *y* in Program 10.4.

Program 10.4: Checking for goal position within a certain margin in C

```
1    if (abs(GOALX - x) < 50 && abs(GOALY - y) < 50)
2    { LCDSetPrintf(3,0, "Goal found        ");
3      VWSetSpeed(0, 0); // Stop robot
4      return 0;          // Program finished
5    }
```

Afterwards, we distinguish three states: *DRIVING* (starting state), *ROTATING* and *FOLLOWING*. Program 10.5 shows the code for *DRIVING*.

Program 10.5: State "driving" in C

```
1    switch (state)
2    { case DRIVING: // Drive towards the goal
3      if (dists[180]<400 || dists[150]<300 || dists[210]<300)
4      { VWSetSpeed(0, 0); // stop
5        ...
6        state = ROTATING;
7      } else if (abs(diff) > 1.0) VWSetSpeed(200, diff);
8             else VWSetSpeed(200, 0);
9      break;
```

We check whether there is an obstacle in front to avoid a collision. Lidar direction 180° is straight ahead (same as *PSD_FRONT*); directions 150° and 210° are 30° to either side. If there is an obstacle in this range, the robot will stop and go into *ROTATION* mode (Program 10.6).

If there is no obstacle and the angle difference towards the goal is within 1°, the robot will drive straight; otherwise, it will drive a curve.

Instead of doing a more complex wall-following operation, here we simply rotate the robot 90° to the left of the obstacle (within 5° of error), and then drive a small distance away from the obstacle. This behavior can be clearly seen in the screenshots in Figure 10.1.

Program 10.6: State "rotating" in C

```
1    case ROTATING: // Rotate perpendicular to obstacle
2      diff = round(phi - perp);
3      if (abs(diff) > 5) VWSetSpeed(0, 50);
4      else { VWSetSpeed(0, 0);
5            ...
6            state = FOLLOWING;
7         }
8      break;
```

After that, the robot goes into *FOLLOWING* mode, where we need to check whether the robot is back at the last *hit* point, which would mean that there is no path to the goal and it needs to give up (Program 10.7). We use a counter to ensure that the robot has moved significantly away from the *hit* point before it can be recorded as a new position. This is one of the algorithm additions required to make it work in a realistic robot setting in the presence of small errors and noise.

Program 10.7: State "following" in C

```
1    case FOLLOWING: // Follow along obstacle boundary
2      counter++;
3      if (counter>10 && abs(hit_x-x)<50 && abs(hit_y-y)<50)
4      { VWSetSpeed(0, 0);
5        LCDSetPrintf(3,0, "Goal unreachable ");
6        return 1;  // finish with error
7      }
```

For the *leave* condition (terminating wall-following behavior), we need to calculate the values for the shortest distance to the goal so far (*d_min*) and the free space in the goal direction (*f*). The *leave* condition is

$$d - f <= d\_min - STEP$$

If this is the case, then the free space in the goal direction from the current position (*d*) brings us closer towards the goal by a margin (*STEP*) than the present best distance (*d_min*). This will terminate the object following (*leave point*) and the algorithm will restart at step 1, driving directly towards the goal (see Program 10.8).

Program 10.8: Calculating the leave condition in C

```
1   int dx = GOALX - x;
2   int dy = GOALY - y;
3   float d = sqrt(dx*dx + dy*dy);
4
5   // Update minimum distance
6   if (d < d_min) d_min = d;
7
8   // Calculate free space towards goal
9   int angle = 180 - (theta-phi);
10  if (angle<0) angle +=360;
11  int f = dists[angle];
12  LCDSetPrintf(2,0,"a=%d d=%f f=%d m=%f ", angle,d,f,d_min);
13
14  // Check leave condition
15  if (d - f <= d_min - STEP)
16  { VWSetSpeed(0, 0);
17    VWStraight(300, 100);
18    VWWait();
19
20    LCDSetPrintf(3,0, "Leaving obstacle    ");
21    diff = round(theta - phi);
22    VWTurn(diff, 50);
23    VWWait();
24    state = DRIVING
25  }
```

The screenshots in Figure 10.1 and Figure 10.2 show the development of the robot's pathfinding when circumnavigating two obstacles.

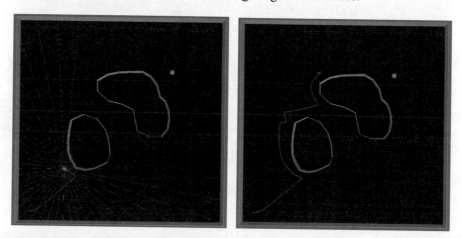

Figure 10.1: DistBug search developing

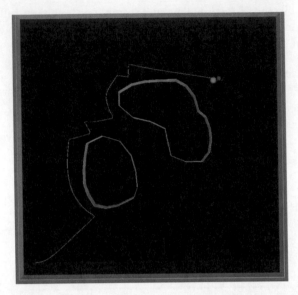

Figure 10.2: Goal found in DistBug algorithm

10.3 Navigation in Known Environments

If we already have a map of our robot's environment, it would be a waste of time using an algorithm like *DistBug* to find a path to the goal. Instead, we can use the map to plan the shortest path ahead of time or *offline*, *before* we actually start driving. There are different methods of specifying an environment map. Here, we use the simplest possible format, called an *Occupancy Grid*, which is a binary image where each pixel represents a small patch of the environment (e.g. 10cm × 10cm). If a pixel is true (black), it represents an obstacle; if it is false (white), it represents free space. In Figure 10.3, we have put together a few examples of such environment files.

  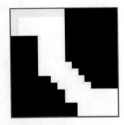

Figure 10.3: Sample occupation grids (black=obstacle, white=free)

10.4 Quadtrees

The Quadtree algorithm takes the environment data and subdivides it into four equal quadrants (labelled counterclockwise 1, 2, 3, 4), as shown in Figure 10.4. Each of the quadrants can be either

- completely free (good to drive through),
- completely blocked (cannot drive there), or
- a mix of free and blocked areas (need to be investigated further).

A tree structure can best represent this algorithm. In the example in Figure 10.4, we have quadrant 1 being completely free and quadrant 3 being completely blocked, while the other two are mixed.

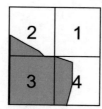 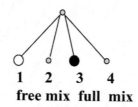

Figure 10.4: Quadtree step 1: four quadrants

Investigating the mixed quadrants further, we call the same algorithm recursively for these nodes. Subdividing only nodes 2 and 4 will give us a more refined tree structure as shown in Figure 10.5. This process is then repeated until each node is completely free or completely blocked, or we have reached the maximum resolution level.

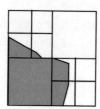 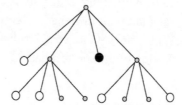

Figure 10.5: Quadtree step 2: further subdivision

Once we have finished the subdivision, we mark all free-space centers as nodes (see Figure 10.6). Some implementations also mark the corners and some side points of the free space squares as nodes, but we will use the simpler version here.

We can easily calculate the coordinates for all these center points a through e, then calculate the path lengths between all nodes. However, not all possible links are actually valid paths; for example, paths c–e and b–e are not valid as they would go through blocked areas, so we have to delete them. The distance graph of the remaining paths in our example is shown in Figure 10.7, right.

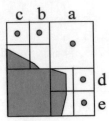
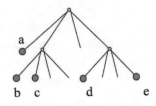

Figure 10.6: Quadtree step 3: free-space centers become nodes

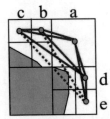
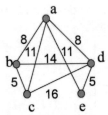

Figure 10.7: Quadtree step 4: collision paths and constructed graph

Assuming the robot is given its starting node and orientation as one of these nodes (otherwise we need to solve the problem of *localization*) plus the goal node position, we can then use a graph algorithm like *Dijkstra's Algorithm* or *A** (A-Star) to find the shortest path [Bräunl 2008][4].

So, in summary, the full navigation algorithm requires a number of steps:

- Quadtree decomposition
- Confirmation that each path is collision-free
- Calculation of the length of each collision-free path
- Application of the A* algorithm for the given start and goal nodes
- Driving of the shortest path

Note the driving problem has now been shifted from a low level (dealing with actual positions) to a high level (working with nodes).

10.5 Quadtree Implementation

The main program will start the quadtree decomposition by calling the recursive subroutine *quad* with the starting position (0,0) and the full size of the map. For simplicity, we use square maps whose size is a power of two, such as

```
quad(0,0, 1024)
```

---

4   T. Bräunl, *Embedded Robotics – Mobile Robot Design and Applications with Embedded Systems*, 3rd Ed., Springer-Verlag, Heidelberg, Berlin, 2008

The recursive routine *quad* in Program 10.9 runs through all the pixels of the given square section, starting at input parameters (x, y) and running in both directions for the given size until $(x+size, y+size)$. The algorithm checks whether every single pixel in this area is free (*false*) or whether every single pixel in this area is occupied (*true*), and then sets the overall variables *allFree* and *allOcc* accordingly.

If either *allFree* or *allOcc* is true, then the subdivision is finished and the procedure *quad* terminates. Only for the case *allFree*, we need to record this position as a free node, e.g., by printing (as in the example) or drawing it; or even better, by entering it into an array for subsequent processing.

If neither *allFree* nor *allOcc* are true, then the area is mixed and we have to further subdivide it. Using $s2 = size/2$ as an abbreviation, we split the area (x,y) into four quadrants of half the side length, starting at

$$(x, y), \ (x+s2, y), \ (x, y+s2), \ (x+s2, y+s2).$$

Program 10.9: Quadtree decomposition in C

```
1   void quad(int x, int y, int size) // start pos + size
2   { bool allFree=true, allOcc = true;
3     for (int i=x; i<x+size; i++)
4       for (int j=y; j<y+size; j++)
5         if (field[i][j]) allFree=false; //at least 1 occ.
6                     else allOcc=false;  //at least 1 free
7     if (allFree) printf("free %d %d (%d)\n", x, y, size);
8       else if (!allOcc  &&  (size>1))
9       { int s2 = size/2;
10        quad(x,     y,     s2);
11        quad(x+s2,  y,     s2);
12        quad(x,     y+s2,  s2);
13        quad(x+s2,  y+s2,  s2);
14      }
15  }
```

This routine will print all free areas found:

```
free    0    0 ( 32)
free   32    0 ( 32)
free    0   32 (  4)
free    4   32 (  4)
free    0   36 (  1)
free    1   36 (  1)
free    2   36 (  2)
free    4   36 (  2)
...
```

Of course, it is nicer to display them graphically on the LCD using *LCDArea*, as we have done in Figure 10.8, right. To declutter the image, we only display free squares with a minimum size of 16×16 pixels.

In order to generate paths, we should store the center points of the free squares in an array for further processing. For generating all possible paths, we

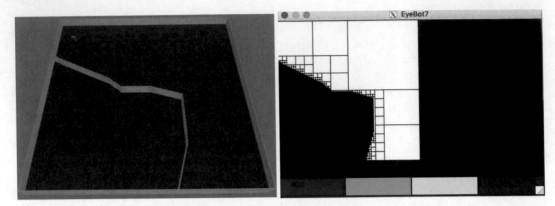

Figure 10.8: Sample environment (left) and its quadtree decomposition (right)

just draw a line from the center of every free square to every other center (see red lines in Figure 10.9).

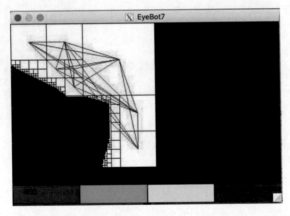

Figure 10.9: Quadtree decomposition with all paths (some of which are intersecting occupied areas)

As can be seen, some of the paths between free squares intersect "forbidden" occupied areas. We have to eliminate these before we can apply a shortest path algorithm. For this, it helps to look at the general case of how a line segment can intersect a square, see Figure 10.10.

The intersection algorithm given by [Alejo 2008][5] is as follows:

1. Are all four corners of box *RSTU* on the same side of line segment *AB*?
   If yes, we are done → no intersection!
   For each corner point $P \in \{R, S, T, U\}$ calculate the line equation for the line going through points *A* and *B*:

5   Alejo, *How to test if a line segment intersects an axis-aligned rectangle in 2D*, 2008, https://stackoverflow.com/questions/99353/how-to-test-if-a-line-segment-intersects-an-axis-aligned-rectangle-in-2d

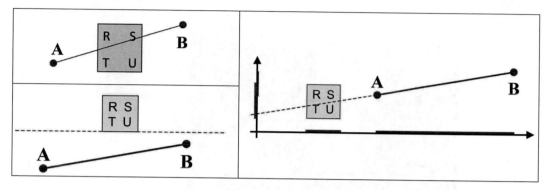

Figure 10.10: Line-box intersection algorithm: check if all corners are on same side of line (bottom left) and check x,y-shadows for overlap (right)

$$F(P) = (B_y - A_y) \cdot P_x + (A_x - B_x) \cdot P_y + (B_x \cdot A_y - A_x \cdot B_y)$$

- $F(P) = 0$ means point P is on the line AB
- $F(P) > 0$ means point P is above line AB
- $F(P) < 0$ means point P is below line AB

2. Are all four Fs positive or are all four Fs negative ?
   If yes, we are done → no intersection!
   If not, project the AB line endpoints onto the x-axis and check if the line shadow intersects the square shadow; then do the same with the y-axis.

   - If $(A_x > U_x$ and $B_x > U_x)$ then there is no intersection
   - If $(A_x < R_x$ and $B_x < R_x)$ then there is no intersection
   - If $(A_y > U_y$ and $B_y > U_y)$ then there is no intersection
   - If $(A_y < R_y$ and $B_y < R_y)$ then there is no intersection
   - Else there is an intersection

Applying this test for each generated line segment (linking each free cube center with each of the others) will eliminate a number of line segments. In the graph shown in Figure 10.11, we have drawn the collision-free paths in blue, while the intersecting paths are shown in red.

The actual distance calculation is simple, as we have the center coordinates for all free squares. The distance between two square centers is given by the Euclidean formula:

$$distance - \sqrt{[(x_2 - x_1)^2 + (y_2 - y_1)^2]}$$

So, the only remaining steps for driving the shortest distance are applying the A* algorithm and actually executing the driving commands.

The sample environments and map files in Figure 10.12 and in previous figures of this chapter have been created by Joel Frewin at the UWA Robotics & Automation Lab.

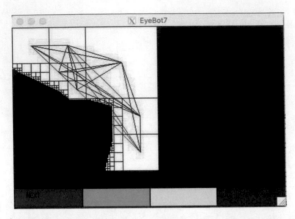

Figure 10.11: Quadtree decomposition with non-intersecting paths

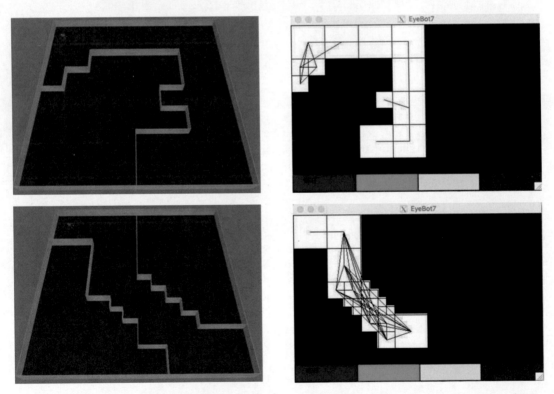

Figure 10.12: More environments with quadtree decompositions and paths

10.6 Shortest Path Algorithm

To find the shortest path in a given distance graph, we can use either Dijkstra's algorithm [Dijkstra 1959][6] or the A* (A-star) algorithm [Hart, Nilsson, Raphael 1968][7]. Dijkstra finds the shortest path from a start node to all other nodes, while A* only finds the shortest path from a specific start node to a specific goal node. The A* algorithm can only be used if the minimum distance between two nodes (usually the Euclidean distance or "air distance") is known. This additional information makes the A* algorithm a lot more efficient in most practical applications.

In the A* example in Figure 10.13, we have five nodes, including the start and goal nodes. Actual distances between the nodes are given along the edges that link them, and we also have the minimum distance to the goal from each node written inside each box (this serves as a *lower bound*, which is required by this algorithm). For example, driving from the start node to node *a* takes 10 meters and we know that the start node itself is at least seven meters from the goal.

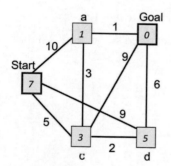

Figure 10.13: Distance graph in initial state

The algorithm explores all choices from the start node and gives them a distance score. So we have the following possible paths:

- S → a   score: 10 (path length) + 1 (min. remainder from *a*) = 11
- S → c   score: 5 + 3 = 8
- S → d   score: 9 + 5 = 14

Sorting these three paths gives us

- **S → c   score: 8**
- S → a   score: 11
- S → d   score: 14

[6]   E. Dijkstra, *A note on two problems in connexion with graphs*, Numerische Mathematik, Springer-Verlag, Heidelberg, vol. 1, pp. 269-271 (3), 1959
[7]   P. Hart, N. Nilsson, B. Raphael, *A Formal Basis for the Heuristic Determination of Minimum Cost Paths*, IEEE Transactions on Systems Science and Cybernetics, vol. SSC-4, no. 2, 1968, pp. 100-107 (8)

z

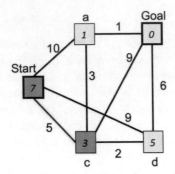

Figure 10.14: Distance graph after selecting sub-path Start–c

Being a *best first search* algorithm, only the shortest path identified so far (S→c shown in red in Figure 10.14) is explored further. Expanding S→c in the next iteration of the A* algorithm gives us three new paths:

- S → c → a   score: $5 + 3 + 1 = 9$
- S → c → b   score: $5 + 9 + 0 = 14$
- S → c → d   score: $5 + 2 + 5 = 12$

Sorting all old and new paths identified so far results in this list:

- **S → c → a   score: 9**
- S → a   score: 11
- S → c → d   score: 12
- S → c → b   score: 14
- S → d   score: 14

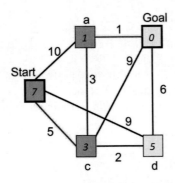

Figure 10.15: Distance graph after selecting sub-path Start–c–a

The shortest path S→c→a is highlighted in red in Figure 10.15 and will be expanded further. There is only one possible path extension:

- **S → c → a → Goal   score: 9**

As this extended path has reached the goal and still has the lowest recorded score, the A* algorithm terminates. We have found the shortest possible path!

10.7 Tasks

- Implement the Quadtree decomposition.
- Implement drivable path detection.
- Implement path length calculation.
- Implement the A* algorithm for finding the shortest distance.
- Implement the driving algorithm to combine all of the above.

ROBOT VISION

V ision is quite likely the most important sensor in robotics. Although it requires a lot more computing power and algorithm development than a Lidar sensor in order to extract environment data, it has the great advantage that it is so much cheaper and also provides us with color and intensity data instead of just a distance value.

11.1 Camera and Screen Functions

Each of our real and simulated robots is equipped with a digital camera. For the real robots, this is the Raspberry Pi camera; for the simulated robots it is a virtual camera sensor with similar characteristics in aperture and resolution.

The first task will be to continuously read an image from the camera and then display it unchanged on the LCD. In order to do this, we need to initialize the camera to tell it which image resolution it should use. This initialization also automatically sets the image display size for the LCD functions and it sets the image size for the image processing library functions, which we will look at later. The choices for camera resolution are

- QQVGA   160 ×   120 pixels
- QVGA   320 ×   240 pixels
- VGA   640 ×   480 pixels
- CAM1MP   1,296 ×   730 pixels
- CAMHD   1,920 × 1,080 pixels
- CAM5MP   2,592 × 1,944 pixels

As a first step, it is probably a good idea to select a moderate image size. We choose QVGA, as this image resolution still fits completely onto the on-board LCD we are using.

We also filled up the scenery with some objects from the simulator's drop down menu. Figure 11.1 shows the robot environment with the corresponding camera image from the robot's point of view displayed on its LCD.

© Springer Nature Switzerland AG 2020
T. Bräunl, *Robot Adventures in Python and C*, https://doi.org/10.1007/978-3-030-38897-3_11

 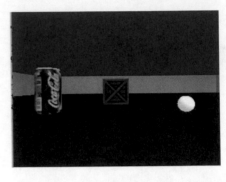

Figure 11.1: Robot with objects (left) and local camera view (right)

We already had a look at programs to read and display a camera image at the beginning of this book, but here are the minimal versions. First, we present an endless loop Python program (Program 11.1), next is the equivalent in C (Program 11.2).

Program 11.1: Minimal camera program in Python

```
1   from eye import *
2
3   CAMInit(QVGA)
4   while True:
5     img = CAMGet()
6     LCDImage(img)
```

Program 11.2: Minimal camera program in C

```
1    #include "eyebot.h"
2
3    int main()
4    { BYTE img[QVGA_SIZE];
5
6      CAMInit(QVGA);
7      while (1)
8      { CAMGet(img);
9        LCDImage(img);
10     }
11   }
```

	−1	
−1	4	−1
	−1	

11.2 Edge Detection

After reading an image, we want to extract some meaningful information from the image data. We can do this by

- Programming routines directly in Python, C or C++
- Using the small RoBIOS image processing library
- Using the comprehensive OpenCV library or some other library [Kaehler, Bradski 2017][1]

Laplace template

	−1	
−1	4	−1
	−1	

As an example, we use edge detection [Bräunl et al. 2001][2], which tries to find the outlines of objects (and shadows) in an image by finding grayscale discontinuities. The simplest to implement (but not the best in quality) is the Laplace function shown on the left, which takes a grayscale image as input. For every pixel, the edge value is calculated as four times its grayscale value minus the four neighboring pixel values (up, down, left, right). So, in a uniform section of an image, all five values will be similar and the function value will be close to zero. However, if the pixel is at the border between a darker and a brighter area, the function value will be a high positive or negative number – indicating an *edge*.

Note, that in most grayscale images, each pixel is given as a byte value, so this gives us a value range of [0, 255]. In this scheme, 0 means black, 255 means white and all other values are shades of gray.

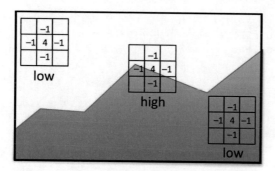

Figure 11.2: Sample image with some Laplace template locations

In the sample image in Figure 11.2 we assume a bright upper part and a dark lower part. Going from pixel to pixel over the whole image lets us see the different scenarios. If all of the pixels in the template are bright (top left filter placement) or all of them are dark (bottom right), then the Laplace filter output value will be very low. However, if some template pixels are bright and some are dark, then we will get a high positive or a high negative value depending on whether the current pixel is brighter than its surroundings or darker. For the

---

[1] A. Kaehler, G. Bradski, *Learning OpenCV 3: Computer Vision in C++ with the OpenCV Library*, O'Reilly, 2017
[2] T. Bräunl, S. Feyrer, W. Rapf, M. Reinhardt, *Parallel Image Processing*, Springer Verlag, Heidelberg Berlin, 2001

three sample areas in Figure 11.2 we calculate the function output values in a simplified environment as follows:

- top left: $4 \cdot 255 - 255 - 255 - 255 - 255 \quad = \quad \mathbf{0}$
- middle: $4 \cdot 255 - 255 \quad -0 - 255 \quad -0 \quad = \mathbf{510}$
- bottom right: $4 \cdot \quad 0 \quad -0 \quad -0 \quad -0 \quad -0 \quad = \quad \mathbf{0}$

A high absolute value indicates an edge (transition from dark to bright area or vice versa); a low absolute value indicates no edge (uniform brightness area).

The C function in Program 11.3 is mainly a loop through all pixels, applying the Laplace template. The loop does not run over the full range [0, *width*height*]. Instead, we start at *width* and stop at *width*(height–1)* as we have to avoid any access outside of the array boundaries.

If the current pixel is referenced by *i*, then the pixel at the left will be at position *i–1* and the one to the right at *i+1*; the pixel above will be at position *i–width* and the one below at *i+width*. Since the subtraction of two byte values [0, 255] can easily result in an integer outside these boundaries, we use the absolute function followed by a test for constant 255 to ensure that each result value will be in the range [0, 255].

Program 11.3: Laplace filter programmed without library functions in C

```
1   #include "eyebot.h"
2
3   void Laplace(BYTE gray_in[], BYTE gray_out[])
4   { int i, delta;
5
6     for (i = IP_WIDTH; i < (IP_HEIGHT-1)*IP_WIDTH; i++)
7     { delta  = abs(4 * gray_in[i]
8                 -gray_in[i-1]        - gray_in[i+1]
9                 -gray_in[i-IP_WIDTH] - gray_in[i+IP_WIDTH]);
10      if (delta > 255) delta = 255;
11      gray_out[i] = (BYTE) delta;
12    }
13  }
14
15  int main()
16  { BYTE img[QVGA_PIXELS], lap[QVGA_PIXELS];
17
18    CAMInit(QVGA);
19    while (1)
20    { CAMGetGray(img);
21      Laplace(img, lap);
22      LCDImageGray(lap);
23    }
24  }
```

We did not calculate the output values that are left out by the *for*-loop (first and last row), so these values should be set to zero (black). Also, note that we

	−1	
−1	4	−1
	−1	

use a single loop and do not take the row format into account. So, the Laplace filter for the rightmost pixel in a row will also use the leftmost pixel of the next row in its calculations. Because of this, we should disregard a 1-pixel wide border around the resulting image. The output in Figure 11.3 now gives us the object outlines in white on a black background, as expected.

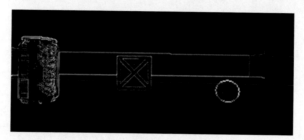

Figure 11.3: Laplace filter output

Using the built-in *IPLaplace* function makes life considerably simpler and the output will be identical. The following two programs show the implementation in Python (Program 11.4) and in C (Program 11.5). The Python solution is, as always, more compact and follows a slightly different parameter syntax. We return image values as a result in Python, for example

```
edge = IPLaplace(gray)
```

while we use them as a second (output) parameter in C:

```
IPLaplace(img, edge);
```

Program 11.4: Laplace filter program using RoBIOS library in Python

```
1   from eye import *
2
3   CAMInit(QVGA)
4   while True:
5     gray = CAMGetGray()
6     edge = IPLaplace(gray)
7     LCDImageGray(edge)
```

A nice extension is to use the Laplace result as a color overlay on the original gray image. Function *IPOverlayGray* adds the second gray image over the first one, using the color given (here *RED*). Since the resulting Laplace output *lap* is not just black and white pixels, there are many gray values in between, we need to add a threshold function as an intermediate step. This weeds out all the "weak edges" (in this case a value less than 50). Program 11.6 contains the complete code and Figure 11.4 shows the output of the program.

The same application is shown in Python in Program 11.7. The complete list of the RoBIOS image processing functions is shown in Figure 11.5. This is only a small number of functions, but they implement the most frequently used features.

Program 11.5: Laplace filter program using RoBIOS library in C

```
1    #include "eyebot.h"
2
3    int main()
4    { BYTE img[QVGA_PIXELS], edge[QVGA_PIXELS];
5
6      CAMInit(QVGA);
7      while (1)
8      { CAMGetGray(img);
9        IPLaplace(img, edge);
10       LCDImageGray(edge);
11     }
12   }
```

Program 11.6: Laplace filter with result overlaid onto input image in C

```
1    #include "eyebot.h"
2
3    void Threshold(BYTE gray[])
4    { int i;
5      for (i=0; i<QVGA_PIXELS; i++)
6        gray[i] = (gray[i] > 50);
7    }
8
9    int main()
10   { BYTE img[QVGA_PIXELS], lap[QVGA_PIXELS], col[QVGA_SIZE];
11
12     CAMInit(QVGA);
13     while (1)
14     { CAMGetGray(img);
15       IPLaplace(img, lap);
16       Threshold(lap);
17       IPOverlayGray(img, lap, RED, col);
18       LCDImage(col);
19     }
20   }
```

Figure 11.4: Laplace edges in red overlaid onto original image

	-1	
-1	4	-1
	-1	

Program 11.7: Laplace filter with result overlaid onto input image in Python

```
1    from eye import *
2
3    def Threshold(gray):
4      for i in range(0, QVGA_PIXELS):
5        if (gray[i] > 50):
6          gray[i]=255
7        else:
8          gray[i]=0
9
10   CAMInit(QVGA)
11   while True:
12     gray = CAMGetGray()
13     edge = IPLaplace(gray)
14     Threshold(edge)
15     col = IPOverlayGray(gray, edge, RED)
16     LCDImage(col)
```

```
int    IPSetSize(int resolution);                              // Set IP resolution
int    IPReadFile(char *filename, BYTE* img);                  // Read PNM file
int    IPWriteFile(char *filename, BYTE* img);                 // Write color file
int    IPWriteFileGray(char *filename, BYTE* gray);            // Write gray file
void   IPLaplace(BYTE* grayIn, BYTE* grayOut);                 // Laplace edges
void   IPSobel(BYTE* grayIn, BYTE* grayOut);                   // Sobel edge
void   IPCol2Gray(BYTE* imgIn, BYTE* grayOut);                 // color to gray
void   IPGray2Col(BYTE* imgIn, BYTE* colOut);                  // gray to color
void   IPRGB2Col (BYTE* r, BYTE* g, BYTE* b, BYTE* imgOut);    // 3*gray to col
void   IPCol2HSI (BYTE* img, BYTE* h, BYTE* s, BYTE* i);       // RGB to HSI
void   IPOverlay(BYTE* c1, BYTE* c2, BYTE* cOut);              // Overlay col
void   IPOverlayGray(BYTE* g1, BYTE* g2, COLOR col, BYTE* cOut); // Ov. gray
COLOR  IPPRGB2Col(BYTE r, BYTE g, BYTE b);                     // RGB to color
void   IPPCol2RGB(COLOR col, BYTE* r, BYTE* g, BYTE* b);       // color to RGB
void   IPPCol2HSI(COLOR c, BYTE* h, BYTE* s, BYTE* i);         // RGB to HSI
BYTE   IPPRGB2Hue(BYTE r, BYTE g, BYTE b);                     // RGB to hue
void   IPPRGB2HSI(BYTE r, BYTE g, BYTE b, BYTE* h, BYTE* s, BYTE* i); // hue
```

Figure 11.5: Image processing library functions

11.3 OpenCV

Starting as an Intel Research project in 1999, OpenCV[3] is today one of the most frequently used image processing libraries. It has bindings for Python, C++ and Java, and supports advanced AI tools such as TensorFlow and Caffe.

Luckily, OpenCV uses the same file format as EyeBot/EyeSim/RoBIOS for color and grayscale images. We only need the conversion function *frame* to

---

[3] OpenCV weblink, https://opencv.org

transfer an image across as OpenCV stores additional information, such as width and height, together with the image data in one structure. Displaying an OpenCV image with a RoBIOS function can then be done by just accessing the raw data with *myimage.data*. Program 11.8 shows an example using OpenCV functions in C++.

Program 11.8: Canny edge detection using OpenCV library functions in C++

```
1   #include "opencv2/highgui/highgui.hpp"
2   #include "opencv2/imgproc/imgproc.hpp"
3   #include "eyebot++.h"
4   using namespace cv;
5
6   int main()
7   { QVGAcol img;
8     Mat edges;
9
10    CAMInit(QVGA);
11    while(1)
12    { CAMGet(img);                                // Get image
13      Mat frame(240, 320, CV_8UC3, img);  // OpenCV conv.
14      cvtColor(frame, edges, COLOR_RGB2GRAY); // RGB-->GRAY
15      GaussianBlur(edges, edges, Size(7, 7), 1.5, 1.5);
16      Canny(edges, edges, 50, 100, 3);    // Canny edge
17      LCDImageGray(edges.data);           // Display result
18    }
19  }
```

The Canny edge detector is a more complex filter and usually gives better results than the simple Laplace function we implemented before (see Figure 11.6).

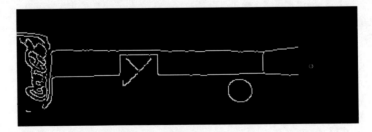

Figure 11.6: Canny edge detection output

11.4 Color Detection

Detecting objects by color is quite simple, provided that only the desired object has the specified color in an image. In comparison, a detection by shape is a lot more complex and requires some mathematical background. In this section, we want to detect a *red ball* as an example.

	-1	
-1	4	-1
	-1	

Most image sensors represent pixels as 3-byte values in RGB format (red, green, blue). So, black is (0,0,0), white is (255,255,255) and "full" red is (255,0,0). Values with three identical components, such as (50,50,50), are shades of gray, while colors with different components are shades of the predominant color.

As there is always some noise in an image, we *cannot* simply check whether a pixel is red, by asking

```
if (r==255 && g==0 && b==0)
```

And as ambient lighting changes all the time, we *cannot* use this more forgiving relation either

```
if (r>200 && g<50 && b<50)
```

Imagine, for the sake of argument, that our "red" pixel in a sunny outdoor scenario has an RGB value of

(210, 20, 10).

If now a cloud covers the sun and the ambient lighting goes down by 50%, then the same pixel will suddenly have the RGB value of

(105, 10, 5).

All component values are being divided by two, so a simple comparison will not work. The solution for this problem is to transfer RGB values to another color space, such as HSI (hue, saturation, intensity), see [Bräunl et al. 2001][4]. In this format

- *hue* [0, 255]   specifies the color value as a position on a circular color *rainbow*.
- *saturation* [0, 255]   specifies the color strength – the lower the saturation, the higher the white component.
- *intensity* [0, 255]   specifies the overall brightness – the lower the intensity, the higher the black component.

In HSI, *white* is {*,0,255) and *black* is (*,*,0). Grayscales are (*,0,*g*), where *g* varies in range 0 (black) to 255 (white). The asterisk "*" is a "don't care term" and stands for any arbitrary value.

Conversion from RGB to HSI is trivial for *S* and *I* but requires trigonometric functions for *H*, which is the component we are most interested in. The following formulas are adapted from [Hearn, Baker, Carithers 2010][5]:

- $I = (R+G*B) / 3$
- $S = 255 - min(R,G,B) / I$
- $H = cos^{-1}[0.5*(R-G + R-B) / \sqrt{((R-G)^2 + (R-B)*(G-B))}]$

We usually use a simplified approximation formula for calculating the hue and we provide the whole transformation as a RoBIOS function *IPCol2HSI*, which makes the transformation a lot easier. As the hue is only a single-byte value, we can now convert an RGB color image to a hue image, which looks like a grayscale image with only one byte per pixel.

---

[4] T. Bräunl, S. Feyrer, W. Rapf, M. Reinhardt, *Parallel Image Processing*, Springer Verlag, Heidelberg Berlin, 2001

[5] D. Hearn, P. Baker, W. Carithers, *Computer Graphics with Open GL*, 4th Ed., Pearson, 2010

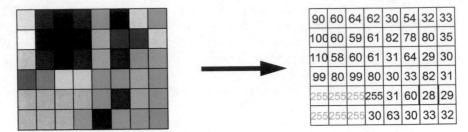

90	60	64	62	30	54	32	33
100	60	59	61	82	78	80	35
110	58	60	61	31	64	29	30
99	80	99	80	30	33	82	31
255	255	255	255	31	60	28	29
255	255	255	30	63	30	33	32

Figure 11.7: RGB color image (left) and hue image (right)

If intensity or saturation are too low, then it will be impossible to assign a proper color hue. You can notice this when using your digital camera in low light environments, when you suddenly get incorrect stray color pixels. We labeled these values with 255, meaning "*no hue*" and will exclude them from subsequent processing.

As can be seen from the sample image in Figure 11.7, each RGB value has now been transferred into a hue value. Values around 60 represent red, which is the color we want to detect in this example. Therefore, the next step will be to specify a hue range for the detection process, e.g., we are looking for a hue in the range [55, 65]. Every hue pixel in this value range will become *true* (1); every pixel outside this range becomes *false* (0). This gives us the binary matching image in Figure 11.8[6].

0	1	1	1	0	0	0	0
0	1	1	1	0	0	0	0
0	1	1	1	0	1	0	0
0	0	0	0	0	0	0	0
0	0	0	0	0	1	0	0
0	0	0	0	1	0	0	0

Figure 11.8: Binary hue match image

We can already see the 1s as a cluster or blob pattern, but we still need to find an algorithmic method to determine the center of our colored object. A very simple and effective method is creating a histogram over all rows and a histogram over all columns. This sounds a lot more complex than it actually is. All we have to do is to add up all the values for each column of the hue-match image and then do the same for each row. As can be seen in the example, for the first row we get

$$0+1+1+1+0+0+0+0 = 3$$

and for the first column we get

$$0+0+0+0+0+0 = 0 \ .$$

---

[6] Note that the hue range could extend to either side of 0, e.g., [250, 5]. This needs special consideration and also needs to leave out our no-hue value of 255.

	-1	
-1	4	-1
	-1	

We enter these values and those for all remaining rows and columns in an additional vector (one-dimensional array) as shown in Figure 11.9.

0	1	1	1	0	0	0	0	**3**
0	1	1	1	0	0	0	0	**3**
0	1	1	1	0	1	0	0	**4**
0	0	0	0	0	0	0	0	**0**
0	0	0	0	0	1	0	0	**1**
0	0	0	0	1	0	0	0	**1**

0	3	3	3	1	2	0	0

Figure 11.9: Binary hue match image with column and row histograms

Now we are almost there. We know in which rows and columns there are a lot of our desired object pixels (high values) and in which there are few or no matching pixels at all (low or zero values). Since the desired object is a round ball, we can find its center by just determining the maximum value of the line and column histogram vector. In the example in Figure 11.9, the maximum value for the row histogram is four and occurs in row three (counting from the top, starting at one). The maximum value of the column histogram is three and this value occurs several times – we just take the first occurrence, which is column two. Therefore, our calculated object center is at (x,y)-position (2,3) counted from the top-left.

Looking at the gray cluster of 1-values in the hue-match image above (as well as in the original color image), one would probably have placed the object center at (x,y)-position (3,2) instead, but remember: we are only one pixel off the "correct" solution, which can always happen in image processing, and we are just looking at a very small sample image of 6×8 pixels. Detection will be better in a full-size image as we will show shortly.

In order to get there, we first implement the hue matching and histogram generation in C. Here, we only generate the column histogram as a simplification. After all, if we want a robot to drive towards an object, we only need to know its x-position and this is what the column histogram will give us.

Program 11.9 runs a loop over all columns (*for-x*) and a nested loop over all rows (*for-y*). If the difference between the hue of the current pixel and the desired hue is below a threshold, then the histogram for the current column is incremented ($hist[x]++$)

The remaining step is to find the maximum in this histogram, which is quite simple. In Program 11.10 we run a single loop (*for-i*) over all elements of the histogram and remember the highest value. Note that at the end of this function, we do *not* return *val*, the highest value found, but instead we return its position *pos*, as we are only interested in the *location* of the colored object.

Program 11.9: Histogram generation in C

```
1    void GenHist(VGAcol img, int hue, line hist, int thres)
2    { int x,y, pos, diff;
3      for (x=0; x<CAMWIDTH; x++)
4      { hist[x] = 0;
5        for (y=0; y<CAMHEIGHT; y++)
6        { pos  = y*CAMWIDTH + x;
7          diff = abs(img[pos] - hue);
8          if ( ((diff < thres) || (255-diff < thres))
9               && (img[pos] != NO_HUE))
10            hist[x]++;
11       }
12     }
13   }
```

Program 11.10: Finding maximum in histogram in C

```
1    int FindMax(line hist)
2    { int i, pos=0, val=hist[0]; // init
3      for (i=1; i<CAMWIDTH; i++)
4        if (hist[i] > val)
5        { pos = i;
6          val = hist[i];
7        }
8      return pos;
9    }
```

The returned value *pos* will be in the range of [0, *CAMWIDTH*–1], which can be easily translated into a steering command for the robot. Value 0 should steer maximum left, *CAMWIDTH*/2 straight ahead and *CAMWIDTH* –1 to the maximum right. Note that a result of 0 could also mean that no matching pixels have been found. In this case, the robot should execute another search command, e.g., rotating on the spot or driving straight until an obstacle is encountered.

Program 11.11: Driving robot according to histogram output in C

```
1    if (pos < CAMWIDTH/3) VWTurn(10, 30);                // left
2      else { if (pos > 2*CAMWIDTH/3) VWTurn(-10, 30); // ri.
3             else VWStraight( 50, 100);                // straight
4           }
```

We don't even have to be that specific regarding the steering angle. In practice, a simple three-way selection statement is all that is required. Turn left, if the object center is in the left third of the image, turn right if it is in the right third, otherwise drive straight (see Program 11.11).

	−1	
−1	4	−1
	−1	

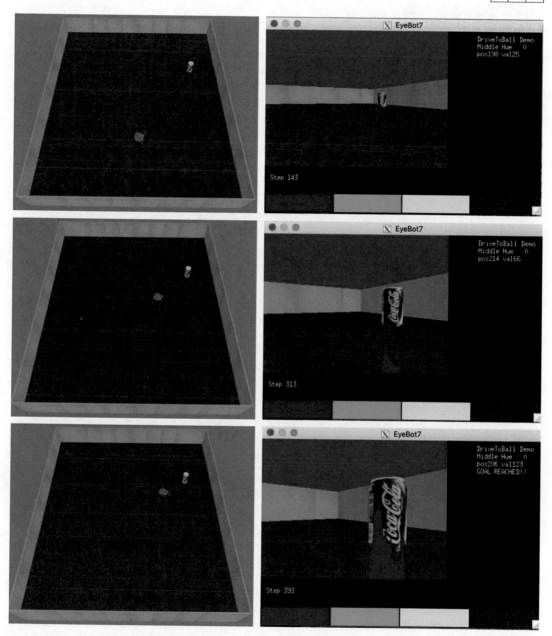

Figure 11.10: Robot driving towards red object with histogram over image

The execution result can be seen in the image sequence in Figure 11.10. Since we run the image detection and drive commands in a continuous loop, the robot will continuously correct its driving angle and will hone in on the red target. We used a can instead of a ball in this example and the white text on the red object clearly affects the histogram, as can be seen in the screenshots. However, the algorithm is still robust enough to let the robot find the red can.

11.5 Motion Detection

Motion detection sounds complicated, but as we will show here, it is really very simple; in fact, it is even simpler than color detection. We want a robot to detect any motion in its visual field. It should then either rotate its camera (if mounted on a servo) or orientate itself towards the center of the motion. Driving towards detected motion while sensing can be tricky, however, as when the robot is in motion, every pixel in its visual field seems to be moving towards an outer edge (check out the opening sequence from *Star Trek*).

Anyway, let us just try to detect motion for a stationary observer robot. The way we do this is to take two camera images, one shortly after the other, and then compare them by subtracting one image from another, pixel by pixel. We use the absolute value of the difference, as we are only interested in pixels that changed between the first and the second image. We do not distinguish whether a pixel gets brighter or darker.

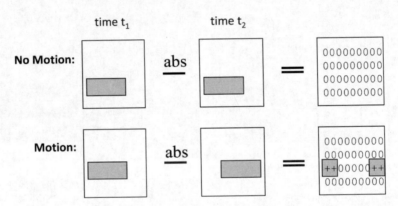

Figure 11.11: Motion detection principle interpreting images as 2D matrices

In the diagram in Figure 11.11, we demonstrate this principle. In the first case, there has been no motion between the first and second image; therefore, we have – except for noise – two largely identical images at times t_1 and t_2. Subtracting these image matrices from each other gives us an image resembling the zero matrix.

However, if there was some motion in the image like the gray block shifting from left to right in the second example, we will get one or more areas in the difference image with a high value. If we then calculate the average difference value over all pixels and compare it with a fixed threshold, we can either say *"yes, there was motion"* or *"no, there was no motion"* in our visual field.

The functions for *image_diff* and *avg* are mostly self-explanatory and are shown in Python (Program 11.12) and C (Program 11.13). Note that for the array declaration in Python, we have to specify the values of type *c_byte* for compatibility reasons, as the EyeBot Python library functions ultimately call the C library.

	−1	
−1	4	−1
	−1	

Program 11.12: Calculating the image difference in Python

```
1    from eye import *
2    from ctypes import *
3
4    def image_diff(i1, i2):
5        diff = (c_byte * QVGA_PIXELS)()
6        for i in range(QVGA_PIXELS):
7            diff[i] = abs(i1[i] - i2[i])
8        return diff

9
10   def avg(d):
11       sum=0
12       for i in range(QVGA_PIXELS):
13           sum += d[i]
14       return int(sum/QVGA_PIXELS)
```

Program 11.13: Calculating the image difference in C

```
1    void image_diff(BYTE i1[SIZE], BYTE i2[SIZE],
2                    BYTE d[SIZE])
3    { for (int i=0; i<SIZE; i++)
4        d[i] = abs(i1[i] - i2[i]);
5    }

6
7    int avg(BYTE d[SIZE])
8    { int i, sum=0;
9      for (i=0; i<SIZE; i++)
10       sum += d[i];
11     return sum / SIZE;
12   }
```

The main function in Program 11.14 for Python (and Program 11.15 for C) first reads two images at 100ms apart, then calls *image_diff*, which is displayed on the screen, followed by *avg*. We print the average difference value to the screen and – if it exceeds a threshold – raise an alarm.

In the execution example, we let one robot continuously rotate on the spot between two cans (by using *VWSetSpeed(0,100)* followed by an endless *while*-loop) while the analyzing robot is watching the scene using the code discussed before. The *SIM* script file is given in Program 11.16.

As can be seen from the screenshot in Figure 11.12, the analyzing robot clearly sees the motion of the other moving robot, but none of the stationary surrounding objects show up in the difference image.

In a second step, we can use a similar technique as before for the color detection and run the same algorithm three times, each time on one third of the input image pair (left, center, right). This will give us a true/false motion result for each of the three sectors and can easily select the sector (if any) in which

Program 11.14: Main program for motion detection in Python

```
1   def main():
2       CAMInit(RES)
3
4       while True:
5           image1 = CAMGetGray()
6           OSWait(100)   # Wait 0.1s
7           image2 = CAMGetGray()
8           diff = image_diff(image1, image2)
9           LCDImageGray(diff)
10          avg_diff = avg(diff)
11          LCDSetPrintf(0,50, "Avg = %3d", avg_diff)
12          if (avg_diff > 15):  # Alarm threshold
13              LCDSetPrintf(2,50, "ALARM!!!")
14          else:
15              LCDSetPrintf(2,50, "        ") # clear text
```

Program 11.15: Main program for motion detection in C

```
1   int main()
2   { BYTE image1[SIZE], image2[SIZE], diff[SIZE];
3     int avg_diff, delay;
4
5     CAMInit(RES);
6     while (1)
7     { CAMGetGray(image1);
8       OSWait(100); // Wait 0.1s
9       CAMGetGray(image2);
10      image_diff(image1, image2, diff);
11      LCDImageGray(diff);
12      avg_diff = avg(diff);
13      LCDSetPrintf(0,50, "Avg = %3d", avg_diff);
14      if (avg_diff > 15) LCDSetPrintf(2,50, "ALARM!!!");
15    }
16  }
```

Program 11.16: Motion detection SIM script

```
1   # Environment
2   world ../../worlds/small/Soccer1998.wld
3   can   300  600  0
4   can   300 1000 45
5
6   # Robot position (x, y, phi) and executable
7   S4 300 800   0 turn.x
8   S4 800 800 180 motion.x
```

	−1	
−1	4	−1
	−1	

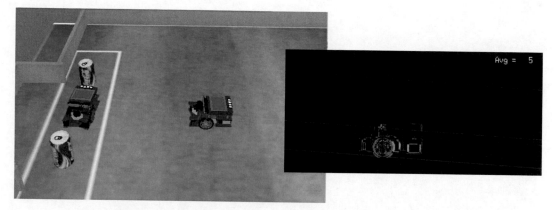

Figure 11.12: Sample scenario with rotating robot and stationary cans (left) and calculated motion image (right)

motion has occurred (see Figure 11.13). If there was motion in more than one sector, we can go back to the average distance values before thresholding and compare them to find out which sector has the highest motion activity.

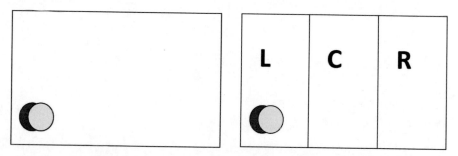

Figure 11.13: Motion in corner of visual field (left) and 3-way decomposition of image (right)

11.6 Tasks

- Write a program that starts from one environment corner, searches for a red can, drives towards it, surrounds it and then pushes it back to its home position.

- Extend the previous program by a function that teaches a desired color hue at a button press. That way, objects of any color can be searched.

- Try to combine motion detection with driving towards the motion location. This requires determining and subtracting the image differences caused by the robot's egomotion.

STARMAN

12

We would now like to introduce a simple articulated walking robot by the name of *Starman*, which was first described by Ngo and Marks in 1994 [Fukunaga et al. 1994] [1].

Starman, as described in the original report, was only a virtual 2D robot, but it was a great tool for experimenting with learning algorithms. As it turns out, moving forward is not that easy, not even for a circular 2D robot with five limbs that cannot fall over.

We went a few steps further, to bring Starman into the real 3D world. We built a physical Starman robot with powerful servos for each of its five legs around a cylindrical body – and then generated its virtual counterpart in Eye-Sim. Each limb can be moved individually via a servo command *SERVOSet*.

Figure 12.1: Real Starman robot versus simulated Starman

[1]   A. Fukunaga, L. Hsu, P. Reiss, A. Shuman, J. Christensen, J. Marks, J. Ngo, *Motion-Synthesis Techniques for 2D Articulated Figures*, Harvard Computer Science Group Technical Report TR-05-94, 1994

T. Bräunl, *Robot Adventures in Python and C*, https://doi.org/10.1007/978-3-030-38897-3_12

12.1 Moving Limbs

The screenshot in Figure 12.1 shows Starman in its initial neutral position. All limbs are in their middle position – 128 of the servo movement range [0, 255]. Moving individual limbs will be our first step (see Figure 12.2). To do this, we place Starman in to our default box environment (Program 12.1).

Program 12.1: Starman SIM script

```
1   robot ../../robots/Articulated/starman.robi
2   Starman 1000 300 1000 90 move.x
```

The move function lets us iterate through all five limbs over the first key, while *KEY2* and *KEY3* let us move the selected limb up or down. Each servo position is stored in the array *pos* and each servo starts in the middle position of 128. We also print the current servo settings to the LCD for each button press (see Figure 12.2, right, and Program 12.2).

Figure 12.2: Starman's limb movement real (left) and simulated (right)

With this, we can generate any Starman configuration, but this is not fast enough to make it walk. Therefore, we wrote a program that only moves a single support limb (limb number three) back and forth. Taking advantage of body mass and friction, this movement will let Starman slowly "shuffle" to the left (Program 12.3 and Figure 12.3).

Moving just one of Starman's support legs will not result in a convincing walking motion – an economic, speedy motion requires coordinated action of several limbs and the optimal sequence cannot be found easily. We could use trial and error to go through different movement patterns, hoping that Starman will finally walk, or we can use an AI method, such as Genetic Algorithms. If we want to try this, we first need to come up with a motion model for Starman.

Program 12.2: Selecting and moving individual limbs in C

```
1    #include "eyebot.h"
2    #define MIN(a,b)  (((a)<(b))?(a):(b))
3    #define MAX(a,b)  (((a)>(b))?(a):(b))
4
5    int main()
6    { int pos[5] = {128,128,128,128,128};
7      int i,k, leg=0;
8
9      LCDMenu("Leg+", "+", "-", "END");
10     do
11     { switch(k=KEYGet())
12       { case KEY1: leg = (leg+1)%5; break;
13         case KEY2: pos[leg] = MIN(pos[leg]+5, 255); break;
14         case KEY3: pos[leg] = MAX(pos[leg]-5,   0); break;;
15       }
16       for (i=0; i<5; i++)
17       { LCDPrintf("S%d pos %d, ", i+1, pos[i]);
18         SERVOSet(i+1, pos[i]);
19       }
20       LCDPrintf("\n");
21     } while (k != KEY4);
22   }
```

Program 12.3: Moving single limb forward/backward to shuffle Starman in C

```
1    #include "eyebot.h"
2
3    int main()
4    { int x, y, phi;
5      for (int i = 0; i < 10; i++)
6      { VWGetPosition(&x, &y, &phi);
7        LCDPrintf("x=%4d, y=%4d\n", x, y);
8        SERVOSet(3, 128+27); OSWait(1000);
9        SERVOSet(3, 128);    OSWait(1000);
10     }
11   }
```

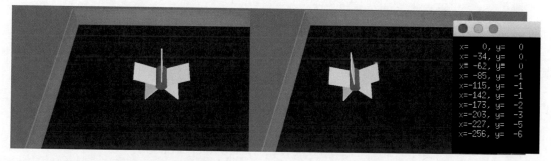

Figure 12.3: Shuffling Starman with position data

12.2 Motion Model

In order to approach things more systematically, we want to establish a motion model for Starman. For this, we assume that Starman's motion pattern will be repetitive, so we will only need to find the motion sequences for each of the five limbs for a relatively short time, e.g. two seconds, then the sequence will repeat. Since we do not want two limbs to collide and they are only 360°/5 = 60° apart, we need to limit their motion to [−30°, +30°] or somewhat less, considering the limb thickness. So, our complete motion solution would look something like the graph in Figure 12.4, where each limb is represented by one curve. Each curve (servo) starts and finishes in the neutral position 0, representing the limb's straight orientation.

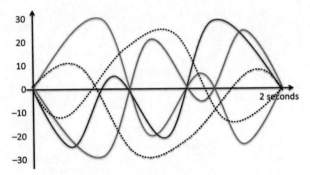

Figure 12.4: Starman motion sequence for five limbs

But we don't know yet what each actual curve function should be. We'll leave this issue until later and first digitize this graph by representing each curve by a number of control points. Using around 10 points per curve (limb movement) should be enough. Movements between control points will either be smoothed automatically by the motor hardware or through code in software. The graph in Figure 12.5 shows the 10 control points for one of the five limbs.

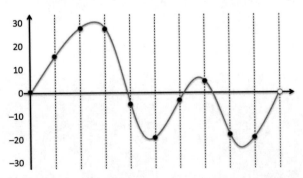

Figure 12.5: Starman control points for a single limb

So this solution would be the series of integer numbers
[0, 15, 28, 27, −5, −20, −4, 6, −20, −21] .

Since we are dealing with a real physical system, it will probably be sufficient to use integer values in the range [−30, +30] rather than rational numbers, so we can easily represent each control point with 1 byte. A full Starman configuration for any point in time would then be 5 bytes (1 byte per limb) and a full motion solution would be 50 bytes (10 control points · 5 bytes).

Details for this approach on a more complex robot can be seen in [Boeing, Bräunl 2015][2].

12.3 Genetic Algorithms

Genetic Algorithms (GAs) are an optimization method for problems that are otherwise hard to solve. GAs maintain a set (*generation*) of individuals (*chromosomes* or *genotypes*), which are encoded as byte sequences. Each chromosome's performance is evaluated by the *fitness* function in the real or simulated world, which will assign a performance value to each chromosome. This value will then determine each chromosome's chance of getting selected for genetic recombination for the next generation. The whole iteration process will stop after a maximum number of generations or when a sufficiently good solution has been found (Figure 12.6).

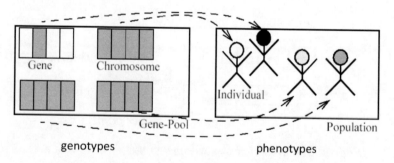

genotypes   phenotypes

Figure 12.6: Generating individuals from chromosomes

Program 12.4 shows this central loop in the main program – terminated either by a maximum number of iterations or when a certain walking performance has been achieved.

In every iteration, we evaluate each individual chromosome by calling the function *evaluate*. The best performing chromosome will be copied unchanged into the next generation. For all other *n* chromosomes, we conduct *n–1 select-gene* operations and with the help of the *crossover* function generate two new chromosomes from each pair of old chromosomes. As a final step, a number of mutations are executed.

2 A. Boeing, T. Bräunl, *Dynamic Balancing of Mobile Robots in Simulation and Real Environments*, in Dynamic Balancing of Mechanisms and Synthesizing of Parallel Robots, Dan Zhang, Bin Wei (Eds.), Springer International, Cham, Switzerland, Dec. 2015, pp. 457-474 (18)

Program 12.4: Gene pool definition and main genetic algorithm loop in C

```
1    BYTE pool[POP][SIZE],
2        next[POP][SIZE];
3    ...
4    for (iter=0; iter<MAX_ITER && maxfit<FIT_GOAL; iter++)
5      { evaluate();
6        memcpy(next[0], pool[maxpos], SIZE); // pres. best
7        for (pos=1; pos<POP; pos+=2)
8        { s1 = selectgene();                 // select 1st
9          s2 = selectgene();                 // select 2nd
10         crossover(s1,s2, pos);             // mating
11       }
12       for (int m=0; m<MUT; m++)  mutation(); // mutations
13       memcpy(pool, next, POP*SIZE);       // copy genepool
14     }
```

Crossover (Figure 12.7 and Program 12.5) is a very simple operation. Taking two chromosomes *A* and *B* that have been selected in the previous step, a random cutting position through their bit-string is determined. Then the left half of chromosome *A* is glued to the right half of chromosome *B* and vice versa. These two new child chromosomes will enter the next generation of the process.

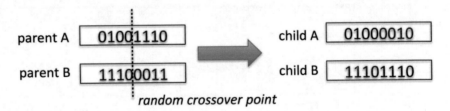

Figure 12.7: Genetic crossover principle

Program 12.5: Crossover function in C

```
1    void crossover(int g1, int g2, int pos)
2    { int cut = rand()%(SIZE-1) +1; // range [1, SIZE-1]
3      memcpy(next[pos ], pool[g1], cut);
4      memcpy(next[pos ]+cut, pool[g2]+cut, SIZE-cut);
5      memcpy(next[pos+1], pool[g2], cut);
6      memcpy(next[pos+1]+cut, pool[g1]+cut, SIZE-cut);
7    }
```

The mutation operation (Figure 12.8 and Program 12.6) is equally simple. A random position in a random chromosome of the current generation is determined to just flip this one bit (0→1 or 1→0). The idea behind this measure is to ensure that the whole search volume is being explored. If, for example, all chromosomes in the current generation started with a 0-bit, then a solution starting with a 1-bit would never be considered if there was no mutation.

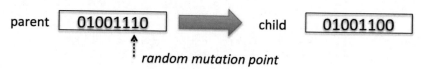

Figure 12.8: Genetic mutation principle

Program 12.6: Mutation function in C

```
1   void mutation()
2   { int ind = rand() % (POP-1) + 1; // [1, POP-1]
3     int pos = rand() % SIZE;
4     int bit = rand() % 8;
5     next[ind][pos] ^= (1<<bit);      // XOR: flip bit
6   }
```

Before starting the evaluation process, we need to initialize our chromosomes with some random values (within reason). We set all limb angle control points to the straight middle position (value 128) and then add random values to it, which will account for ±5.

The limbs of the simulated robot also have to be initialized before we can start. We do this with five calls to the *SERVOSet* function (Program 12.7).

Program 12.7: Initializing gene pool and set function for limbs in C

```
1   void init()
2   { int i, leg, point, pos,val;
3     for (i=0; i<POP; i++)
4     { pool[i][0]=128; pool[i][1]=128; pool[i][2]=128;
5       pool[i][3]=128; pool[i][4]=128; // neutral init
6       for (point=1; point<CPOINTS; point++)
7         for (leg=0; leg<5; leg++)
8         { pos = 5*point+leg;
9           val = pool[i][pos-5] + (rand() & 10) - 5;//r. +/-5
10          pool[i][pos] = MAX(0, MIN(255, val));
11        }
12    }
13  }

14
15  void set(chrom c, int pos)
16  { int leg;
17    for (leg=0; leg<5; leg++) SERVOSet(leg+1, c[5*pos+leg]);
18  }
```

Program 12.8: Fitness function calculation in C

```
1   int fitness(int i)
2   { int rep, point, x, y, phi;
3
4     SIMSetRobot(1, 1000, 1000, 0, 90);
5     VWSetPosition(0,0,0);
6
7     set(pool[i],0);   // starting position
8     OSWait(2000);
9     for (rep=0; rep<REP; rep++)
10      for (point=0; point<CPOINTS; point++)
11      { set(pool[i], point);
12        OSWait(250); // ms
13      }
14    VWGetPosition(&x, &y, &phi);
15    return 1 + abs(x) + abs(y);   // min. fitness 1
16  }
17
18  void evaluate()
19  { fitsum = 0.0;
20    maxfit = 0.0;
21    for (int i=0; i<POP; i++)
22    { fitlist[i] = fitness(i);
23      fitsum    += fitlist[i];
24      if (fitlist[i]>maxfit)  // record max fitness
25      { maxfit=fitlist[i]; maxpos=i; }
26    }
27  }
```

The fitness function in Program 12.8 simply runs the robot through the control points for a given chromosome for a certain number of iterations – then checks in what position the robot has ended up. The further away the robot lands from the starting point, the "fitter" it is.

As the robot wanders around a fair bit after thousands or millions of chromosome evaluations, we need to set the robot back to the same starting point before each simulation run. Otherwise, it might run into a wall or worse, fall off the virtual table. For this, we can use the simulation-only function *SIMSetRobot*.

Function *evaluate* calls the fitness evaluation for each chromosome in the gene pool and stores results in the global variable *fitlist*. It also adds up the total fitness sum of all the chromosomes in the whole generation, which we will need later for the selection process.

The selection function is called twice in every iteration of the main program. It needs to select a random chromosome; however, the random function needs to be biased to reflect each chromosome's fitness. If chromosome *A* has twice the fitness value of chromosome *B*, then *A* should be twice as likely to be selected than *B*.

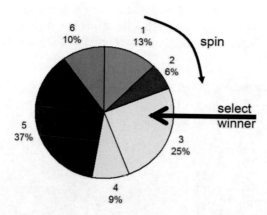

Figure 12.9: Gene (parent) selection principle with "wheel of fitness"

Program 12.9: Gene (parent) selection function in C

```
1   int selectgene()
2   { int  i, wheel,count;
3
4     wheel = rand() % fitsum; // range [0, fitsum-1]
5     i=0;
6     count = fitlist[0];
7     while (count < wheel)
8     {  i++;
9        count += fitlist[i];
10    }
11    return i;
12  }
```

We solve the selection problem with what we call the *"wheel of fitness"* (see Figure 12.9 and Program 12.9). Just imagine the selection process as spinning the wheel on a game show. Wheel segments are covered with chromosome symbols, where the segment size matches the relative fitness level (twice the fitness means twice the area). So, assuming each spin is random, it will select a random chromosome with the desired bias according to their fitness levels.

12.4 Evolution Run

Running the GA program will take a long time. Each chromosome bit-string of the population is evaluated, which runs in real time at around 10 seconds per robot. So, evolving a population of 100 robots over 100 generations will take 10^5 seconds, which equals roughly 28 hours. Of course, this could be executed significantly faster on a more powerful computer system, running the simulator in a "headless mode" with a faster-than-real execution time.

The screenshots in Figure 12.10 show the fitness levels of a population of 15 Starmen after 10 (left) and 80 (right) iterations. The fitness value for the best individual has more than doubled during this evolution process (from 78 to 128).

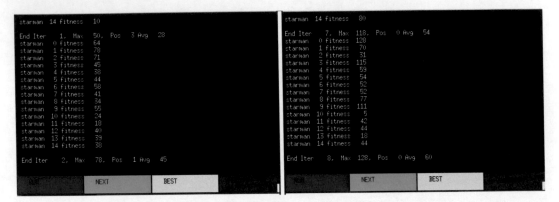

Figure 12.10: Evolution run stages

Finally in Figure 12.11, Starman is shown at its initial position, indicated by a green floor marker (left), and at its final position after executing the evolved gait (right).

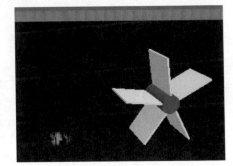

Figure 12.11: Before and after execution of the evolved gait

12.5 Tasks

- Optimize the GA implementation for Starman and find the optimal walking pattern.
- Add an initial start sequence that will lead into the repetitive motion sequence and evolve this also with a GA.
- Extend (or even evolve) Starman into a more complex articulated creature.

DRIVERLESS CARS

13

I n this chapter, we come close to autonomous cars – only at a smaller scale. We use the same robots as before, but we place them in a miniature traffic scenario, complete with lane markings, traffic signs, parking areas and other cars (or robots). Even pedestrians (figurines), houses, trees and more can be included.

13.1 Autonomous Model Car Competitions

There are at least two student competitions in this area, the Carolo-Cup[1], an annual event organized by the Technical University Braunschweig, Germany, and the very similar Audi Autonomous Driving Cup[2], organized by the car manufacturer Audi. While the Carolo-Cup is directly open to all participants, Audi Cup participants first need to pass a preselection phase. The successful teams then receive a free autonomous model car for the competition.

In the following, we concentrate on the Carolo-Cup, which we scaled down for our robots at a ratio of 2:1. We rebuilt the standard Carolo intersection loop on a large table in our lab as well as in simulation (see Figure 13.1).

The Carolo-Cup allows us to develop and improve on a large range of tasks relevant for automotive research. All of the algorithms developed for the small robot cars can then be extended for applications in real autonomous vehicles. The areas covered are

- Lane detection and lane keeping
- Collision avoidance
- Detection of other vehicles and pedestrians
- Traffic sign recognition
- Automated parking
- Automated overtaking

---

[1] TU Braunschweig, *Carolo-Cup*, https://wiki.ifr.ing.tu-bs.dc/carolocup/carolo-cup
[2] Audi AG, *Audi Autonomous Driving Cup*, https://www.audi-autonomous-driving-cup.com

© Springer Nature Switzerland AG 2020
T. Bräunl, *Robot Adventures in Python and C*, https://doi.org/10.1007/978-3-030-38897-3_13

Figure 13.1: Real environment versus simulated environment

- Automated intersection control
- Automated zebra crossing detection
- Automated speed control following signage
- Vehicle-to-vehicle and vehicle-to-base-station communication

The beauty of this competition is that new teams can start with just one or two functionalities, e.g., lane keeping and collision avoidance, to develop a working system and then gradually improve their implementation while also adding new functionalities.

13.2 Carolo-Cup

All we need for the Carolo-Cup setup is the standard Carolo loop as an image file from Carolo's website, which we then convert into a real or simulated environment. We complete the scene with some handmade traffic signs, for which there are also image files on the Carolo website.

With a few traffic signs, the Carolo *SIM* script looks like Program 13.1 and will create the environment shown in Figure 13.2. The basic Carolo world file in Program 13.2 (here without walls) is extremely simple, as it just uses the bitmap for the floor.

In the extended example shown in Figure 13.2, we added walls that match exactly our lab setup for the real robots – which also helps to keep the simulated robots from falling off the table.

The software design and implementation work for the Carolo-Cup system described in the following sections was implemented by UWA visiting students Shuangquan Sun, Jingwen Zheng, Zihan Lin (all from the University of Science and Technology of China), Zihan Qiao and Shanqi Liu (both from Zhejiang University).

Program 13.1: Carolo-Cup SIM script

```
1    # Environment
2    world ../../worlds/small/Carolo.wld
3
4    # Objects
5    object ./objects/ParkingSign/ParkingSign.esObj
6    object ./objects/SpeedLimitSign/cancelspeedlimitsign.esObj
7    object ./objects/SpeedLimitSign/speedlimitsign.esObj
8    object ./objects/StopSign/stopsign.esObj
9
10   # Objects
11   ParkingSign 990   223 192
12   StopSign    2270 1192 121
13   StopSign    2301 1922 312
14   CancelSpeedLimitSign 1899 2861 1
15   SpeedLimitSign        46 1820 87
16
17   # robotname x y phi
18   S4 1637 352 180 lane.x
```

Program 13.2: Carolo-Cup world file

```
1    floor_texture carolo-lab.png
2    width  3100
3    height 3100
```

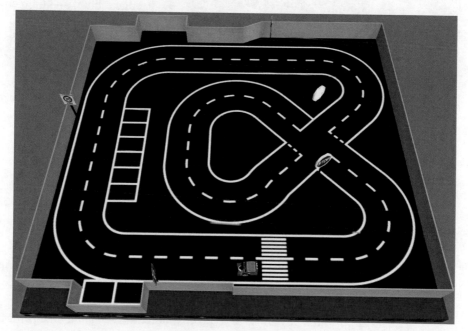

Figure 13.2: Carolo-Cup environment in EyeSim

13.3 Lane Keeping

The first step is conducting image processing to find lane markings, and then using this information to generate a model of the most likely lane curvature. We use OpenCV (see Chapter 11 on robot vision) for all image processing as it is a very versatile and comprehensive library. This also means that our application programs have to be written in C++ or Python, as C is not supported by OpenCV. The processing steps are shown in Figure 13.3.

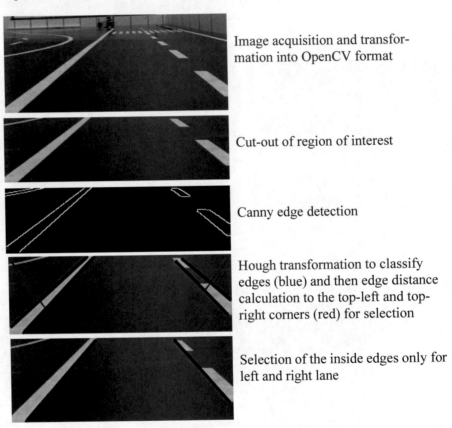

Image acquisition and transformation into OpenCV format

Cut-out of region of interest

Canny edge detection

Hough transformation to classify edges (blue) and then edge distance calculation to the top-left and top-right corners (red) for selection

Selection of the inside edges only for left and right lane

Figure 13.3: Lane detection algorithm steps

It should be noted that many of these operations are very compute-intensive. This is especially the case for the Hough transformation (used here) as well as for feature validation, curve fitting with Lagrange interpolation polynomial and many other operations (not used in this project). The compute power of our real robots is rather limited through the Raspberry Pi controller and we do not want to transmit images to a "remote brain" and receive back driving instructions, which would reduce our robots to remote-controlled cars. Instead, we have an interest in developing fast-computing, alternative vision algorithms that allow us to process automotive vision on an embedded controller.

From the previous steps, we can now calculate the vehicle's relative position to the lane center and correct the curvature of its driving path, to keep it in the middle of the lane. This works quite well for straight passages.

Unfortunately, this method does not work as nicely in a curve. The standard Raspberry Pi camera has a very narrow field of view, so when entering a curve, the robot will only see the outer lane. This problem could of course be fixed in hardware, either by an alternative camera with a wider lens (probably the easiest solution) or by mounting the camera on a servo and then rotating the camera to keep both side lanes in the field of view. We can still try to make sense of a single visible lane as shown below; however, this will be at the cost of a reduced robustness of the program. This means that a slight disturbance can get the autonomous vehicle off track and make it leave the driving lane completely.

Detecting only the left lane marking in a right-hand turn

Detecting only the dashed middle lane marking in a left-hand turn *(remember, we drive on the left side of the road in Australia)*

Figure 13.4: Lane markings in a right/left turn

As can be seen from the sample images in Figure 13.4, there can be very little edge data available in a curve, especially if only the dashed middle lane marking is in view. The less edge pixels we have, the more error-prone the curve detection algorithms will be.

13.4 Intersections and Zebra Crossings

Intersections and zebra crossings are special cases that need to be handled by a Carolo program. In a sense, they are the two extreme cases in terms of lane markings visible in an image frame. At the start of an intersection, there are no vertical lane markings visible at all. At a zebra crossing, there is a larger number (typically around 15) of vertical edges. So, this can be a good first criterion to distinguish these traffic situations.

At an intersection (see Figure 13.5, top), the autonomous vehicle must detect the horizontal stopping line and come to a full stop. It must then check for any cross traffic (a wide-angle lens definitely helps) before driving through the intersection.

At a zebra crossing (see Figure 13.5, bottom), the autonomous vehicle must slow down and check for pedestrians (we use little model figurines) before it can drive through the crossing.

Start of intersection

Zebra crossing
(detected bars marked in blue)

Figure 13.5: Counting vertical edges to distinguish intersections, crossings and regular lanes

13.5 Traffic Sign Recognition

Traffic sign recognition is a lot more complex and we use a combination of learning methods for this as described in [Sun et al. 2019][3]. We use the Histogram of Oriented Gradients (HOG) method for this task [Dalal, Triggs 2005][4], which can run on a Raspberry Pi 3 at a reasonable frame rate of around 3Hz for a QVGA image resolution (320×240 pixels). Figure 13.6 shows some of the signs to be detected and their corresponding HOG patterns.

Figure 13.6: Traffic signs and derived HOG patterns

[3] S. Sun, J. Zheng, Z. Qiao, S. Liu, Z. Lin, T. Bräunl, *Architecture of a driverless robot car based on EyeBot system*, 3rd International Conference on Robotics: Design and Applications (RDA 2019), Xi'an, China, April 2019

[4] N. Dalal, B. Triggs, *Histograms of oriented gradients for human detection,* IEEE Computer Society Conference on Computer Vision and Pattern Recognition (CVPR'05), 2005, pp. 886-93

In Figure 13.7 we show some examples of image frames with superimposed sign recognition. Note that this even works with multiple signs in the same image frame.

Figure 13.7: Traffic sign recognition in driving sequence

The diagram in Figure 13.8 shows the execution time on a Raspberry Pi 3. The average processing time per image frame is 340ms but it varies somewhat. The standard Raspbian operating system does not support real-time operation, but this is also not really necessary, as long as there are no major outliers in image processing time. Traffic sign recognition runs as a separate process on the controller, so it will not affect other control algorithms that have to run at a faster frequency, such as distance sensor evaluation for collision avoidance. The actual PID motor control happens outside the Raspberry Pi on the separate EyeBot-IO controller, which is based on an Atmel processor and communicates via a USB link.

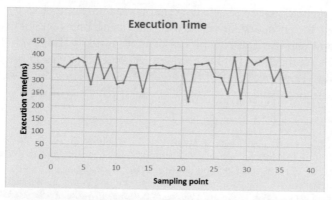

Figure 13.8: Execution timing diagram for traffic sign recognition

The precision (positive prediction value), recall (sensitivity) and F-score for the testing dataset of our implementation are shown in Figure 13.9.

Traffic signs	Precision	Recall	F
Stop	82.61%	61.29%	70.37%
Pedestrian	100%	73.33%	84.62%
Park	100%	50.00%	66.67%
Speed limit	100%	63.64%	77.78%
Speed limit cancel	100%	42.11%	59.26%
Give way	100%	60.00%	75.00%

Figure 13.9: Precision, recall and F1-score for traffic sign recognition

13.6 End-to-End Learning

Neural networks and deep learning gain more and more momentum in autonomous driving software. The approaches presented in this chapter so far range from the traditional engineering approach, where specific image processing functions are applied (e.g. for detecting lane markings), to the use of learning algorithms for a particular task (e.g. traffic sign recognition). However, the ultimate goal for learning systems is *end-to-end* learning. The idea is to present real live data (e.g. a video feed from the driver's perspective) together with the correct desired output (here the steering angle) to a deep neural network and let it learn the complete task in one piece without a programmer having to dissect or preprocess the input data.

Clearly, such an end-to-end system will be a tremendous improvement over traditional AI (Artificial Intelligence) systems and save significant costs in developing a learning system. For autonomous driving, all that is required is a recording of the video feed and steering angle of some good drivers in a large variety of driving scenarios, and the deep neural network will learn how to drive autonomously. Such a scenario was implemented by researchers at Nvidia for an actual drive-by-wire car, using one of their parallel GPGPU (general purpose graphics processor unit) systems [Bojarski et al. 2016][5]. A similar approach, only for general object detection, was implemented at Google Inc. [Howard et al. 2017][6]. Figure 13.10 explains the principle of end-to-end learning.

---

[5] M. Bojarski, D. Del Testa, D. Dworakowski, B. Firner, B. Flepp, P- Goyal, L. Jackel, M. Monfort, U. Muller, J. Zhang, X. Zhang, J. Zhao, K. Zieba, *End to End Learning for Self-Driving Cars*, Nvidia Corporation, Apr. 2016, pp. (9)

[6] A. Howard, M. Zhu, B. Chen, D. Kalenichenko, W. Wang, T. Weyand, M. Andreetto, H. Adam, *MobileNets: Efficient Convolutional Neural Networks for Mobile Vision Applications*, Google Inc., Apr. 2017, pp. (9)

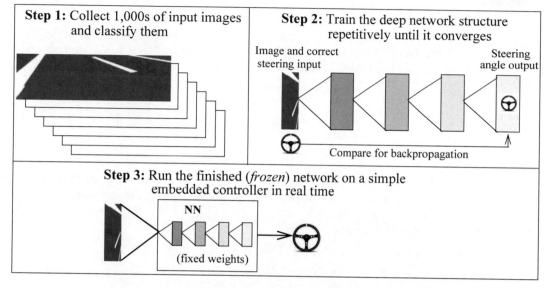

Figure 13.10: End-to-end learning for autonomous driving

Nicholas Burleigh and Jordan King together with the author have used a simplified deep network based on the Nvidia and Google approaches to train a robot driving the Carolo-Cup track and detecting traffic signs [Burleigh, King, Bräunl 2019][7]. We recorded 1,000 test images together with the correct steering angles from running the previously described engineering solution of the Carolo-Cup problem. Figure 13.11 shows sample images for the categories "steer left" (left two images), "drive straight" (middle two images) and "steer right" (right two images).

Figure 13.11: End-to-end learning of traffic scenes for turning left (left), driving straight (middle) and turning right (right)

We then trained a deep neural network using TensorFlow[8] with this data and ten possible steering output values (from full left via straight to full right). The resulting trained network was small enough so it could run on a robot's

---

[7] N. Burleigh, J. King, T. Bräunl, *Deep Learning for Autonomous Driving*, Intl. Conf. on Digital Image Computing: Techniques and Applications (DICTA), Dec. 2019, Perth, pp. (6)

[8] TensorFlow, https://www.tensorflow.org

on-board Raspberry 3 controller with around 9 fps (frames per second) successfully navigating the Carolo-Cup track.

Although simulated and real camera images look similar, the trained network for one system could not be transferred to the other, so it required separate input data collections and separate training sessions for simulated and real robots.

13.7 Tasks

Implement your own version of the Carolo-Cup. Start with one functionality and then add more and more features until you have a comprehensive autonomous driving system:

- Lane detection and lane keeping
- Collision avoidance
- Detection of other vehicles and pedestrians
- Traffic sign recognition
- Automated parking
- Automated overtaking
- Automated intersection control
- Automated zebra crossing detection
- Automated speed control following signage
- Vehicle-to-vehicle and vehicle-to-base-station communication
- Implement an end-to-end learning system for the tasks listed above

FORMULA SAE

14

The Society of Automotive Engineers (SAE) has been conducting the Formula SAE competition for many years. It is an international competition for engineering students to build a one-seater race car from the ground up (this means welding pipes together to form the chassis) and compete in several race and endurance events. In addition to these "dynamic events" there are also "static events", in which teams receive points for the car's design, technology, marketing etc.

Formula SAE and the related Formula Student Germany (FSG) used to be events mainly for mechanical engineering students and were based on petrol cars, often using motorcycle engines. This has changed over the last few years. First was the introduction of an electric vehicle competition, and then in 2017 came the first autonomous vehicle competition [FSG 2016][1].

UWA has been quite successful in the Formula SAE competition, being world champion in 2008. Within the Renewable Energy Vehicle Project (REV) at UWA, we built one of the first SAE Electric cars and one of the first SAE Autonomous cars – both years before there was a competition in these categories.

14.1 Electric Driving

We built our first electric Formula SAE car as part of the REV project in 2010. We inherited a rolling chassis without a drive train from the UWA Motorsport group that competed in the petrol competition. We decided to use a dual motor solution to drive the rear wheels individually from two motor controllers, linked through an electronic differential (see Figure 14.1 and Figure 14.2).

We built a second-generation SAE-Electric vehicle from scratch, designed by Master's student Ian Hooper (Figure 14.2, right). It had four individual

---

[1] Formula Student Germany, *Autonomous Driving at Formula Student Germany 2017,* Aug. 2016, https://www.formulastudent.de/pr/news/details/article/autonomous-driving-at-formula-student-germany-2017/

© Springer Nature Switzerland AG 2020
T. Bräunl, *Robot Adventures in Python and C*, https://doi.org/10.1007/978-3-030-38897-3_14

Figure 14.1: REV Formula SAE Electric 2010 chassis

Figure 14.2: J. Wan, I. Hooper and T. Bräunl with the 1st generation Formula SAE Electric car (left) and wheel hub motors of 2nd generation car (right)

wheel hub motors and more powerful batteries in two side containers. Figure 14.3 shows the original CAD vehicle design from Ian Hooper and its transformation into a virtual robot car in EyeSim.

14.2 Drive by Wire

While an electric drive system certainly helps with implementing an autonomous driving system, it is technically not a prerequisite. The first step for any autonomous vehicle is to implement a drive-by-wire system (see construction by Jordan Kalinowski in Figure 14.4 and Figure 14.5). This means that the car's three main functions: steering, braking and accelerating can be controlled from a computer system. A low-level embedded processor will control these

Figure 14.3: Formula SAE Electric CAD design by Ian Hooper (left) and simulation model in EyeSim (right)

three functions and receive commands from a higher-level intelligent drive computer.

Figure 14.4: REV drive-by-wire control by Jordan Kalinowski

1. Steering

Actuating the steering is probably the hardest part. It typically requires a powerful motor linked to the steering column via a belt drive so that a computer command can change the steering angle (Figure 14.5, left). Of course, this immediately raises safety concerns (which we will address in a safety section later):

- The driver/passenger could get his/her hand stuck in the steering wheel while the drive computer turns it.
- The driver/passenger may want to interfere with the steering but may not be able to physically overpower the motor on the steering column.
- The motor/belt setup may get jammed up or otherwise interfere with the driver's steering operation in manual driving mode.

2. Braking

We did not want to interfere with the friction brakes directly for safety reasons, so we built a lever that can pull the brake pedal from behind, which is driven by a powerful servo. This leaves the driver in control in manual as well as in autonomous mode (Figure 14.5, middle).

3. Accelerating

Computerizing acceleration is actually the simplest part. Since all modern acceleration pedals are already electronic, all it needs is an analog multiplexer that switches between the pedal and the computer output as the input for the motor controller (Figure 14.5, right).

Figure 14.5: Drive-by-wire: steer, brake and accelerate

Figure 14.6: REV Autonomous SAE drive-by-wire and safety system

14.3 Safety System

As a full-size vehicle can be extremely dangerous to the driver/passenger as well as to bystanders, we need to implement several layers of safety systems, both on-board the car and on an external base station. A separate, dedicated embedded controller takes over the safety feature implementation on the vehicle (see Figure 14.6). The safety system and the high-level control system were implemented by Thomas Drage [Drage, Kalinowski, Bräunl 2014][2]:

On-Board

a. Emergency stop buttons
One button for the driver/passenger to disengage autonomous mode and one button to shut down power to the drive system.

b. Electronic heartbeat between car and base station
If the heartbeat is lost, the vehicle will stop.

c. Geofencing
If the car leaves a predefined GPS area, it will stop.

d. Watchdog timer
If the on-board software hangs due to a hardware or software error, the watchdog timer will shut the vehicle power down and apply the brakes via the low-level controller.

e. Manual override
Pressing the accelerator or brake pedal or applying some force on the steering wheel will be detected and revert the vehicle back to manual mode.

Off-Board

a. Remote emergency stop
Stop buttons are implemented on the remote station both as a physical button (linked via USB to the laptop) and as a clickable software button. Engaging either stop buttons will send a stop command to the vehicle.

b. Electronic heartbeat
The counterpart to the vehicle heartbeat.

14.4 Autonomous Driving

For autonomous driving, we need a number of sensors, so that the system can get an accurate account of the vehicle's position, orientation and speed as well as its environment, including all obstacles, other vehicles, pedestrians and so on. From this, it needs to calculate a desired path that will then be sent as steering/braking/acceleration commands to the low-level drive-by-wire controller. The link between the high-level and the low-level controller can be made either via dedicated data lines or via a bus system, such as CAN or USB. Typical sensors for autonomous vehicles such as the UWA/REV cars in Figure 14.7 and Figure 14.8 include

- Lidar (either single or multiple layers)
- Radar
- Camera

---

[2] T. Drage, J. Kalinowski, T. Bräunl, *Integration of Drive-by-Wire with Navigation Control for a Driverless Electric Race Car*, IEEE Intelligent Transportation Systems Magazine, pp. 23-33 (11), Oct. 2014

- Inertial measurement unit (IMU)
- Wheel encoder
- Distance sensor

Figure 14.7: Fully autonomous Formula SAE Electric vehicle with sensors above the driver's seat

We do not want to get into further details on sensor types and sensor operation, but rather look into software implementation for driving tasks. More details can be found in [Lim et al. 2018][3] on Lidar-based autonomous driving and in [Teoh, Bräunl 2012][4] for vision-based autonomous driving. This research was conducted on a donated BMW X5 (see Figure 14.8).

Figure 14.8: UWA/REV autonomous BMW X5 with EyeBot M6 stereo vision controller and windscreen mount

---

[3] K. Lim, T. Drage, R. Podolski, G. Meyer-Lee, S. Evans-Thompson, J. Yao-Tsu Lin, G. Channon, M. Poole, T. Bräunl, *A Modular Software Framework for Autonomous Vehicles*, IEEE Intelligent Vehicles Symposium (IV), 2018, Chang Shu China, pp. 1780–1785 (6)

[4] S. Teoh, T. Bräunl, *Symmetry-Based Monocular Vehicle Detection System*, Journal of Machine Vision and Applications, Springer, vol. 23, no. 4, July 2012, pp. 831–842 (12)

14.5 Cone Track Racing

The 2018 competition rules by FSG/F-SAE require an autonomous car to drive a track outlined by cones with specific distances and colors. As there are no other obstacles and no other vehicles on the track, the easiest method to detect these cones is by using a single-beam Lidar sensor that is mounted horizontally in front of the vehicle at a low height, so it can easily scan all cones in a 180° range.

The competition rules require different colored cones for the left and right side of the track (blue and yellow, respectively), but this does not matter when using a Lidar, as it only reports distance values and not colors. A camera system can be used to supplement cone detection and improve performance and safety – or even be the sole sensor in order to build a cheaper self-driving system.

Figure 14.9: Autonomous cone-track driving on various test circuits

The REV cone-driving algorithms have been implemented by Chao Zhang and team [Lim et al. 2019][5], [Brogle et al. 2019][6] and their results are shown in Figure 14.9.

The first step for simulating this system is to refamiliarize ourselves with EyeSim's Lidar sensor. By default, it reports back 360 distance values, covering a full circle of 360°. However, this can be changed in the robot's *ROBI* description file, e.g., to deliver 1,000 values over a 180° range.

In Program 14.1 we have set up a *SIM* script for our F-SAE car in an empty plane that only contains three orange cones.

We can now click and drag the car around the driving area and observe changes in the image and Lidar data displayed on the LCD (Figure 14.10).

5   K. Lim, T. Drage, C. Zhang, C. Brogle, W Lai, T. Kelliher, M. Adina-Zada, T. Bräunl, *Evolution of a Reliable and Extensible High-Level Control System for an Autonomous Car*, IEEE Transactions on Intelligent Vehicles, 2019, pp. 396–405 (10)
6   C. Brogle, C. Zhang, K. Lim, T. Bräunl, *Hardware-in-the-Loop Autonomous Driving Simulation*, IEEE Transactions on Intelligent Vehicles, 2019, pp. 375–384 (10)

Program 14.1: Formula SAE SIM script

```
1    settings VIS
2
3    # World File
4    world field2.wld
5
6    # Robots
7    robot "../../robots/Ackermann/SAE.robi"
8    SAE 4000 1200 90 conedrive.x
9
10   # Objects
11   object "../../objects/ConeOrange/coneorange.esObj"
12
13   #Left side of the track
14   Cone-O 4000 6500 0
15   Cone-O 3000 6100 0
16   Cone-O 5000 6100 0
```

Figure 14.10: Autonomous cone-track driving on various test circuits

The main function in Program 14.2 for displaying camera and Lidar sensor data is an extension of the Lidar plotting program from Chapter 4. The function *getmax* is a small routine that returns the highest distance value from the current Lidar scan. The function *LIDARSet* sets the Lidar's scanning angle and angular resolution. This could also be specified in the car's *ROBI*-file.

On the screen in Figure 14.10, right, we see the camera image from the car's point of view as well as the 180° Lidar data. As there are no surrounding walls (as you would expect on a race course), all Lidar points not hitting a cone will come back with the maximum value (in this case 9,999mm). Each of the three cones leaves a deep cut in the Lidar diagram, but it is also evident that the middle cone (obstacle) is slightly further away than the two outside ones (the middle black gap starts slightly higher than the two outside ones).

Although the complete system for driving the real Formula SAE car is a lot more complex and has a number of important safety features, we can implement a simplified working algorithm for cone track racing in EyeSim as shown in Program 14.3. The idea implemented here is to use a single-layer forward-

Program 14.2: Formula SAE Lidar demonstration in C

```
 1   int main ()
 2   { int    i, k, m;
 3     int    scan[POINTS];
 4     float  scale;
 5     BYTE   img[QQVGA_SIZE];
 6
 7     LCDMenu("DRIVE", "STOP", "", "END");
 8     CAMInit(QQVGA);
 9     LIDARSet(180, 0, POINTS); // range, tilt, points
10
11     do
12     { k = KEYRead();
13       CAMGet(img);
14       LCDImage(img);
15       if (k==KEY1) VWSetSpeed(200,0);
16       if (k==KEY2) VWSetSpeed(  0,0);
17       LIDARGet(scan);
18       m = getmax(scan);
19       scale = m/150.0;
20       LCDSetPos(13,0);
21       LCDPrintf("max %d scale %3.1f\n", m,scale);
22       for (i=0; i<10; i++)
23         LCDPrintf("%4d ",scan[i*(POINTS/10)]);
24       // plot distances
25       for (i=0; i<PLOT; i++)
26       { LCDLine(180+i,150-scan[SCL*i]/scale,180+i,150,BLUE);
27         LCDLine(180+i,150-scan[SCL*i]/scale,180+i,  0,BLACK);
28       }
29       LCDLine(180+POINTS/(2*SCL),
30               0,180+POINTS/(2*SCL),150, RED);
31     } while (k!=KEY4);
32   }
```

facing Lidar sensor mounted below cone height, which will reliably detect cones in a 180° field of view. As can be seen Figure 14.11, we search for the two most central obstacles detected by the Lidar, left and right of the red middle line (see Figure 14.11, left) in order to find a collision-free steering angle.

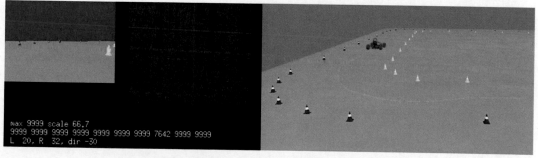

Figure 14.11: SAE car's camera and Lidar view (left) and track driving (right)

The black Lidar shadows in Figure 14.11, left, look almost like the poles in slalom skiing and we use a similar technique to avoid them.

Program 14.3 shows the main iteration of the driving routine, assuming the car is driving with a constant speed set by the function *MOTORDrive*. As the middle of the visual field shifts significantly in a curve, we use variable *middle* when we call auxiliary functions *getleftcone* and *getrightcone*. We then update the middle position between these two cones and calculate the new steering angle as the deviation from the straight direction. Function *SERVOSet* is called for the steering, where the value 128 represents driving straight. Adding the cone gap *dir* to this middle value lets us follow the cone track.

Program 14.3: Main loop of simplified SAE cone track driving in C

```
 1    do  // car is already running with MOTORDrive(1, SPEED)
 2    { OSWait(100);   // reduce main loop to 10Hz
 3      CAMGet(img);
 4      LCDImage(img);
 5
 6      LIDARGet(scan); // Lidar set to 180°range at 180 pt.
 7      m = getmax(scan);
 8      scale = m/150.0;
 9
10      l = getleftcone (scan, middle);  // left-most cone
11      r = getrightcone(scan, middle);  // right-most cone
12      if (l>0 && r>0 && l<r)
13      { middle = (l+r)/2; // middle position of [0..POINTS]
14        dir = (POINTS/2 - middle); // range +/-POINTS/2
15        SERVOSet(1, 128+dir); // 0=right 128=mid. 255=left
16      }
17
18      // plot distances and remove previous line
19      for (i=0; i<PLOT; i++)
20      { LCDLine(180+i,150-scan[SCL*i]/scale,180+i,150,BLUE);
21        LCDLine(180+i,150-scan[SCL*i]/scale,180+i,0,BLACK);
22      }         // draw variable middle line
23      LCDLine(180+middle/SCL,0, 180+middle/SCL,150, RED);
24    } while (k!=KEY4);
25  }
```

The auxiliary functions *getleftcone*, *getrightcone* and *getmax* are listed in Program 14.4. Function *getleftcone* receives the Lidar sensor array together with the current middle position. The function then iterates from this middle position towards the left until it finds the first obstacle – an object closer than 9,000mm. The function will then stop and return the position value. Function *getrightcone* operates in the same fashion, only iterating from the middle towards the right side. Function *getmax* is required for scaling the Lidar data before plotting it to the screen.

Program 14.4: Auxiliary cone track functions in C

```
1    int getleftcone(int a[], int mid)
2    { int i;
3      for (i=mid-1; i>20; i--)
4        if (a[i] < 9000) return i; // cone detected !
5      return -1; // no cone
6    }
7
8    int getrightcone(int a[], int mid)
9    { int i;
10     for (i=mid+1; i<160; i++)
11       if (a[i] < 9000) return i; // cone detected !
12     return -1; // no cone
13   }
14
15   int getmax(int a[])
16   { int i, pos = 0;
17     for (i=1; i<POINTS; i++)
18       if (a[i] > a[pos]) pos = i;
19     return a[pos];
20   }
```

The screenshot in Figure 14.12 now shows the final outcome of the cone-track racing in action. We could very well make this a simulation competition in its own right!

Figure 14.12: Simulated cone track driving of a Formula SAE car

14.6 Tasks

- Implement your version of the cone racing program using only a Lidar sensor.
- Implement your version of the cone racing program using only a camera sensor.
- Implement your version of the cone racing program using a combination of Lidar and camera sensors.
- Extend your program to detect other cars and avoid collisions with them. Race two robot cars (with different programs) against each other in the same circuit.

OUTLOOK

15

If you have completed this book and want to work on larger, more powerful (and more expensive) robot systems, then looking at the Robot Operating System (ROS)[1], originally developed by Willow Garage, should be your next step. ROS is a free open source platform and provides a comprehensive library of high-level robotics software packages and utilities, such as SLAM (Simultaneous Localization and Mapping), visualization (*rviz*), data recording (*rosbag*) and simulation (*Gazebo*). ROS implementations exist for the majority of commercially available research robots and can be used for their program development as well as their simulation.

The drawback of using ROS for beginners is that it requires Ubuntu as the operating system and does not support C as an application language, only C++ and Python. Also, its system structure is significantly more complex and not easy to grasp for robotics novices. We have developed a ROS client for our EyeBot robots and may also include it in EyeSim at a later stage. Our projects involving larger robots and our autonomous vehicles are based on ROS and may migrate to the open hardware/software automotive platform Apollo[2] in the future.

We hope we have inspired you to dive deeper into the world of robotics and carry out many more experiments on your own. The EyeSim simulation environment gives you a chance to develop your robot programs in a realistic, versatile and free environment. On the other hand, we believe it is essential to complete the second step and build a physical robot. This does not have to be expensive, as we have outlined at the beginning of this book. A robot can be built quite cheaply by setting up an embedded controller, like the Raspberry Pi, with a camera, display, two motors and some distance sensors – or alternatively, by converting a remote-controlled model car for less accurate but faster driving.

Have fun and enjoy your robot adventures!

---

[1] Robot Operating System, http://www.ros.org
[2] Apollo Open Platform, http://apollo.auto

© Springer Nature Switzerland AG 2020
T. Bräunl, *Robot Adventures in Python and C*, https://doi.org/10.1007/978-3-030-38897-3_15

APPENDIX

. . . .

RoBIOS-7 Library Functions

Version 7.2, Jan. 2020-- RoBIOS is the operating system for the EyeBot controller.
The following libraries are available for programming the EyeBot controller in C or C++. Unless noted otherwise, return codes are 0 when successful and non-zero if an error has occurred.

In application source files include
```
#include "eyebot.h"
```
Compile application to include RoBIOS library
```
$gccarm myfile.c -o myfile.o
```

- LCD Output
- Key Input
- Camera
- Image Processing
- System Functions
- Timer
- USB/Serial
- Audio
- Distance Sensors
- Servos and Motors
- V-Omega Driving Interface
- Digital and Analog I/O
- IR Remote Control
- Radio Communication
- Multitasking
- Simulation

LCD Output

```
int LCDPrintf(const char *format, ...);            // Print string and arguments on LCD
int LCDSetPrintf(int row, int column, const char *format, ...);  // Printf from given position
int LCDClear(void);                                // Clear the LCD display and display buffers
int LCDSetPos(int row, int column);                // Set cursor position in pixels for subsequent printf
int LCDGetPos(int *row, int *column);              // Read current cursor position
int LCDSetColor(COLOR fg, COLOR bg);               // Set color for subsequent printf
```

© Springer Nature Switzerland AG 2020
T. Bräunl, *Robot Adventures in Python and C*, https://doi.org/10.1007/978-3-030-38897-3

int LCDSetFont(int font, int variation);   // Set font for subsequent print operation
int LCDSetFontSize(int fontsize);   // Set font-size (7..18) for subsequent operation
int LCDSetMode(int mode);   // Set LCD Mode (0=default)
int LCDMenu(char *st1, char *st2, char *st3, char *st4); // Set menu entries for soft buttons
int LCDMenuI(int pos, char *string, COLOR fg, COLOR bg); // Set menu for i-th entry with color
int LCDGetSize(int *x, int *y);   // Get LCD resolution in pixels
int LCDPixel(int x, int y, COLOR col);   // Set one pixel on LCD
COLOR LCDGetPixel (int x, int y);   // Read pixel value from LCD
int LCDLine(int x1, int y1, int x2, int y2, COLOR col); // Draw line
int LCDArea(int x1, int y1, int x2, int y2, COLOR col, int fill); // Draw filled/hollow rectangle
int LCDCircle(int x1, int y1, int size, COLOR col, int fill);   // Draw filled/hollow circle
int LCDImageSize(int t);   // Define image type for LCD (default QVGA;)
int LCDImageStart(int x, int y, int xs, int ys);   // Def. start and size (def. 0,0; max_x, max_y)
int LCDImage(BYTE *img);   // Print color image at screen start pos. and size
int LCDImageGray(BYTE *g);   // Print gray image [0..255] black..white
int LCDImageBinary(BYTE *b);   // Print binary image [0..1] white..black
int LCDRefresh(void);   // Refresh LCD output

Font Names and Variations
HELVETICA (default), TIMES, COURIER
NORMAL (default), BOLD

Color Constants (COLOR is data type "int" in RGB order)
RED (0xFF0000), GREEN (0x00FF00), BLUE (0x0000FF), WHITE (0xFFFFFF), GRAY (0x808080), BLACK (0)
ORANGE, SILVER, LIGHTGRAY, DARKGRAY, NAVY, CYAN, TEAL, MAGENTA, PURPLE, MAROON, YELLOW,
OLIVE

LCD Modes
LCD_BGCOL_TRANSPARENT, LCD_BGCOL_NOTRANSPARENT, LCD_BGCOL_INVERSE,
LCD_BGCOL_NOINVERSE, LCD_FGCOL_INVERSE, LCD_FGCOL_NOINVERSE, LCD_AUTOREFRESH,
LCD_NOAUTOREFRESH, LCD_SCROLLING, LCD_NOSCROLLING, LCD_LINEFEED, LCD_NOLINEFEED,
LCD_SHOWMENU, LCD_HIDEMENU, LCD_LISTMENU, LCD_CLASSICMENU, LCD_FB_ROTATE,
LCD_FB_NOROTATION

Keys

int KEYGet(void);   // Blocking read (and wait) for key press (returns KEY1...KEY4)
int KEYRead(void);   // Non-blocking read of key press (returns NOKEY=0 if no key)
int KEYWait(int key);   // Wait until specified key has been pressed
int KEYGetXY (int *x, int *y);   // Blocking read for touch at any position, returns coordinates
int KEYReadXY(int *x, int *y);   // Non-blocking read for touch at any position, returns coordinates

Key Constants
KEY1...KEY4, ANYKEY, NOKEY

Camera

int CAMInit(int resolution);   // Change camera resolution (incl. IP resolution)
int CAMRelease(void);   // Stops camera stream
int CAMGet(BYTE *buf);   // Read one color camera image
int CAMGetGray(BYTE *buf);   // Read gray scale camera image

For the following functions, the Python API differs slightly as indicated.
```
def CAMGet    () -> POINTER(c_byte):
def CAMGetGray() -> POINTER(c_byte):
```

Resolution Settings
QQVGA(160×120), QVGA(320×240), VGA(640×480), CAM1MP(1296×730), CAMHD(1920×1080),
CAM5MP(2592×1944), CUSTOM (LCD only)
Variables CAMWIDTH, CAMHEIGHT, CAMPIXELS (=width*height) and CAMSIZE (=3*CAMPIXELS) will
be automatically set (BYTE is data type "unsigned char")

Constant sizes in bytes for color images and number of pixels
QQVGA_SIZE, QVGA_SIZE, VGA_SIZE, CAM1MP_SIZE, CAMHD_SIZE, CAM5MP_SIZE
QQVGA_PIXELS, QVGA_PIXELS, VGA_PIXELS, CAM1MP_PIXELS, CAMHD_PIXELS,
CAM5MP_PIXELS

Data Types
```
typedef QQVGAcol  BYTE [120][160][3];        typedef QQVGAgray  BYTE [120][160];
typedef QVGAcol   BYTE [240][320][3];        typedef QVGAgray   BYTE [240][320];
typedef VGAcol    BYTE [480][640][3];        typedef VGAgray    BYTE [480][640];
typedef CAM1MPcol BYTE [730][1296][3];       typedef CAM1MPgray BYTE [730][1296];
typedef CAMHDcol  BYTE[1080][1920][3];       typedef CAMHDgray  BYTE[1080][1920];
typedef CAM5MPcol BYTE[1944][2592][3];       typedef CAM5MPgray BYTE[1944][2592];
```

Image Processing

Basic image processing functions using the previously set camera resolution are included in the RoBIOS
library. For more complex functions see the OpenCV library.

```
int   IPSetSize(int resolution);                    // Set IP resolution using CAM constants
int   IPReadFile(char *filename, BYTE* img);        // Read PNM file, fill/crop; 3:col., 2:gray, 1:bin
int   IPWriteFile(char *filename, BYTE* img);       // Write color PNM file
int   IPWriteFileGray(char *filename, BYTE* gray);  // Write gray scale PGM file
void  IPLaplace(BYTE* grayIn, BYTE* grayOut);       // Laplace edge detection on gray image
void  IPSobel(BYTE* grayIn, BYTE* grayOut);         // Sobel edge detection on gray image
void  IPCol2Gray(BYTE* imgIn, BYTE* grayOut);       // Transfer color to gray
void  IPGray2Col(BYTE* imgIn, BYTE* colOut);        // Transfer gray to color
void  IPRGB2Col (BYTE* r, BYTE* g, BYTE* b, BYTE* imgOut);// Transform 3*gray to color
void  IPCol2HSI (BYTE* img, BYTE* h, BYTE* s, BYTE* i);// Transform RGB image to HSI
void  IPOverlay(BYTE* c1, BYTE* c2, BYTE* cOut);// Overlay c2 onto c1, all color images
void  IPOverlayGray(BYTE* g1, BYTE* g2, COLOR col, BYTE* cOut); // Overlay gray images
COLOR IPPRGB2Col(BYTE r, BYTE g, BYTE b); // PIXEL: RGB to color
void  IPPCol2RGB(COLOR col, BYTE* r, BYTE* g, BYTE* b);// PIXEL: color to RGB
void  IPPCol2HSI(COLOR c, BYTE* h, BYTE* s, BYTE* i);// PIXEL: RGB to HSI for pixel
BYTE  IPPRGB2Hue(BYTE r, BYTE g, BYTE b); // PIXEL: RGB to hue (0 for gray values)
void  IPPRGB2HSI(BYTE r, BYTE g, BYTE b, BYTE* h, BYTE* s, BYTE* i); // PIXEL: RGB to hue
```

For the following functions, the Python API differs slightly as indicated.
```
from typing import List
from ctypes import c_int, c_byte, POINTER

def IPLaplace (grayIn: POINTER(c_byte)) -> POINTER(c_byte):
def IPSobel (grayIn: POINTER(c_byte)) -> POINTER(c_byte):
```

```
def IPCol2Gray (img: POINTER(c_byte)) -> POINTER(c_byte):
def IPCol2HSI (img: POINTER(c_byte)) -> POINTER(c_byte) -> List[c_byte, c_byte, c_byte]:
def IPOverlay (c1: POINTER(c_byte), c2: POINTER(c_byte)) -> POINTER(c_byte):
def IPOverlayGray (g1: POINTER(c_byte), g2: POINTER(c_byte)) -> POINTER(c_byte):
```

System Functions

```
char * OSExecute(char* command);          / Execute Linux program in background
int OSVersion(char* buf);                 // RoBIOS Version
int OSVersionIO(char* buf);               // RoBIOS-IO Board Version
int OSMachineSpeed(void);                 // Speed in MHz
int OSMachineType(void);                  // Machine type
int OSMachineName(char* buf);             // Machine name
int OSMachineID(void);                    // Machine ID derived from MAC address
```

Timer

```
int OSWait(int n);                                    // Wait for n/1000 sec
TIMER OSAttachTimer(int scale, void (*fct)(void));   // Add fct to 1000Hz/scale timer
int OSDetachTimer(TIMER t);                           // Remove fct from 1000Hz/scale timer
int OSGetTime(int *hrs,int *mins,int *secs,int *ticks);  // Get system time (ticks in 1/1000 sec)
int OSGetCount(void);                                 // Count in 1/1000 sec since system start
```

USB/Serial Communication

```
int  SERInit(int interface, int baud,int handshake); // Init communication (see HDT file)
int  SERSendChar(int interface, char ch);       // Send single character
int  SERSend(int interface, char *buf);         // Send string (Null terminated)
char SERReceiveChar(int interface);             // Receive single character
int  SERReceive(int interface, char *buf, int size);  // Receive String (Null term.), returns size
int  SERFlush(int interface);                   // Flush interface buffers
int  SERClose(int interface);                   // Close Interface
```

Communication Parameters
Baudrate: 50 ... 230400
Handshake: NONE, RTSCTS
Interface: 0 (serial port), 1..20 (USB devices, names are assigned via HDT entries)

Audio

```
int AUBeep(void);                    // Play beep sound
int AUPlay(char* filename);          // Play audio sample in background (mp3 or wave)
int AUDone(void);                    // Check if AUPlay has finished
int AUMicrophone(void);              // Return microphone A-to-D sample value
```

Use Analog data functions to record microphone sounds (channel 8).

Distance Sensors

Position Sensitive Devices (PSDs) use infrared beams to measure distance and need to be calibrated in HDT to get correct distance readings. LIDAR (Light Detection and Ranging) is a single-axis rotating laser scanner.

```
int PSDGet(int psd);                          // Read distance value in mm from PSD sensor
int PSDGetRaw(int psd);                        // Read raw value from PSD sensor [1..6]
int LIDARGet(int distance[]);                  // Measure distances in [mm]; def. 360°, 360 points
int LIDARSet(int range, int tilt, int points); // range [1..360°], tilt angle down, number of points
```

PSD Constants
PSD_FRONT, PSD_LEFT, PSD_RIGHT, PSD_BACK
PSD sensors in these directions are connected to ports 1, 2, 3, 4.

LIDAR Constants
LIDAR_POINTS Total number of points returned
LIDAR_RANGE Angular range covered, e.g. 180°

Servos and Motors

Motor and Servo positions can be calibrated through HDT entries.

```
int SERVOSet(int servo, int angle);           // Set servo [1...14] position to [0..255]
int SERVOSetRaw (int servo, int angle);       // Set servo [1...14] position bypassing HDT
int SERVORange(int servo, int low, int high); // Set servo [1...14] limits in 1/100 sec
int MOTORDrive(int motor, int speed);         // Set motor [1...4] speed in percent [-100 ...+100]
int MOTORDriveRaw(int motor, int speed);      // Set motor [1...4] speed bypassing HDT
int MOTORPID(int motor, int p, int i, int d); // Set motor [1...4] PID controller values [1...255]
int MOTORPIDOff(int motor);                   // Stop PID control loop
int MOTORSpeed(int motor, int ticks);         // Set controlled motor speed in ticks/100 sec
int ENCODERRead(int quad);                    // Read quadrature encoder [1...4]
int ENCODERReset(int quad);                   // Set encoder value to 0 [1...4]
```

V-Omega Driving Interface

This is a high-level wheel control for differential driving. It always uses motor 1 (left) and motor 2 (right). Motor spinning directions, motor gearing and vehicle width are set in the HDT file.

```
int VWSetSpeed(int linSpeed, int angSpeed);      // Set fixed linSpeed  [mm/s] and [degrees/s]
int VWGetSpeed(int *linSspeed, int *angSpeed);   // Read current speeds [mm/s] and [degrees/s]
int VWSetPosition(int x, int y, int phi);        // Set robot position to x, y [mm], phi [degrees]
int VWGetPosition(int *x, int *y, int *phi);     // Get robot position as x, y [mm], phi [degrees]
int VWStraight(int dist, int lin_speed);         // Drive straight, dist [mm], lin. speed [mm/s]
int VWTurn(int angle, int ang_speed);            // Turn on spot, angle [deg], ang. speed [degrees/s]
int VWCurve(int dist, int angle, int lin_speed);  // Curve, dist [mm], angle [deg], lin. speed [mm/s]
int VWDrive(int dx, int dy, int lin_speed);      // Drive x[mm] straight and y[mm] left, x>|y|
int VWRemain(void);                              // Return remaining drive distance in [mm]
int VWDone(void);                                // Non-blocking check whether drive is finished (1)
int VWWait(void);                                // Suspend thread until drive operation has finished
int VWStalled(void);                             // Number of stalled motor 0 (none), 1, 2, 3 (both)
```

All VW functions return 0 if OK and 1 if error (e.g. destination unreachable).

Digital and Analog Input/Output

int DIGITALSetup(int io, char direction);   // Set IO line [1...16] to in, out, In pull-up, Jn pull-dn
int DIGITALRead(int io);   // Read and return individual input line [1...16]
int DIGITALReadAll(void);   // Read and return all 16 io lines
int DIGITALWrite(int io, int state);   // Write individual output [1...16] to 0 or 1
int ANALOGRead(int channel);   // Read analog channel [1...8]
int ANALOGVoltage(void);   // Read analog supply voltage in [0.01 Volt]
int ANALOGRecord(int channel, int iterations);   // Record analog data at 1kHz (non-blocking)
int ANALOGTransfer(BYTE* buffer);   // Transfer previously recorded data; returns size

Default for digital lines: [1...8] are input with pull-up, [9...16] are output
Default for analog lines: [0...8] with 0: supply-voltage and 8: microphone
IO settings: i: input, o: output, I: input with pull-up res., J: input with pull-down res

IR Remote Control

These commands allow sending commands to an EyeBot via a standard infrared TV remote (IRTV). IRTV models can be enabled or disabled via a HDT entry. Supported IRTV models are: Chunghop L960E Learn Remote.

int IRTVGet(void);   // Blocking read of IRTV command
int IRTVRead(void);   // Non-blocking read, return 0 if nothing
int IRTVFlush(void);   // Empty IRTV buffers
int IRTVGetStatus(void);   // Checks to see if IRTV is activated (1) or off (0)

Defined Constants for IRTV buttons are:
IRTV_0 ... IRTV_9, IRTV_RED, IRTV_GREEN, IRTV_YELLOW, IRTV_BLUE,
IRTV_LEFT, IRTV_RIGHT, IRTV_UP, IRTV_DOWN, IRTV_OK, IRTV_POWER

Radio Communication

These functions require WiFi modules for each robot, one of them (or an external router) in DHCP mode, all others in slave mode. Radio can be activated/deactivated via an HDT entry. The names of all participating nodes in a network can also be stored in the HDT file.

int RADIOInit(void);   // Start radio communication
int RADIOGetID(void);   // Get own radio ID
int RADIOSend(int id, char* buf);   // Send string (Null terminated) to ID destination
int RADIOReceive(int *id_no, char* buf, int size);   // Read bytes from ID source, returns rec. size
int RADIOCheck(void);   // Check if message is waiting: 0 or 1 (non-block.)
int RADIOStatus(int IDlist[]);   // Returns number of robots (incl. self) and ID list
int RADIORelease(void);   // Terminate radio communication

ID numbers match last byte of robots' IP addresses.

Multitasking

For Multitasking, simply use the pthread functions. A number of multitasking sample programs are included in the demo/MULTI directory.

Simulation *only*

These functions will *only* be available when run in a simulation environment, in order to get ground truth information and to repeat experiments with identical setup.

```
void SIMGetRobot (int id, int *x, int *y, int *z, int *phi);
void SIMSetRobot (int id, int  x, int  y, int  z, int  phi);
void SIMGetObject(int id, int *x, int *y, int *z, int *phi);
void SIMSetObject(int id, int  x, int  y, int  z, int  phi);
int  SIMGetRobotCount()
int  SIMGetObjectCount()
```

id=0 means own robot; id numbers run from 1...n

Thomas Bräunl, Remi Keat and Marcus Pham, 1996-2020

Printed in the United States
by Baker & Taylor Publisher Services